Encyclopaedia of

Nanotechnology

Encyclopaedia of
Nanotechnology

Vol. 1 : Fundamentals and Applications

Dr. Rita Khare

AGROTECH PRESS

JAIPUR ■ NEW DELHI

AGROTECH PRESS

Regd. Off. "Bony Residency", Gate No. 2
Opp. Tilak Public School, Vishweswariya Nagar,
Gopalpura Road, Jaipur - 302018, India
Ph: 0141 - 2761280, Telefax: 2761381
E-mail: agrotechpress@gmail.com

Delhi Office: 102, 1st Floor, Satyam House,
4327/3, Ansari Road, Darya Ganj,
New Delhi - 110002, India
Ph: 011-45652440
visit us: www.agrotechpress.com

Distributed by :
KEYA INTERNATIONAL PRESS LLP
E-mail: keyainternationalpressllp@gmail.com

Published in India, 2024

ISBN : 9789384568566 (10 Vol. Set)

Typeset by Agrotech Press

Printed and bound in India by Replika Press Pvt. Ltd.

Contents

Preface

Nanoscience and nanotechnology are evolving at a rapid pace and there have been a number of scientific and technological advancements in these fields in recent times. Nanotechnology is the engineering of functional systems at the molecular scale. It is the engineering of tiny machines and devices — the projected ability to build things from the bottom up, using techniques and tools being developed today to make complete, highly advanced products. Ultimately, nanotechnology will enable control of matter at the nanometer scale, using mechanochemistry. Shortly after this envisioned molecular machinery expands, it will result in a manufacturing revolution. As globalization leads to an increasing interaction between regions and people of the world, it is important to encourage academic growth in emerging scientific topics such as nanotechnology. Nanotechnology, even though rapidly growing, is still just beginning and already has given rise to numerous novel applications for the solution of many current problems. It encompasses several different fields of science in its purview including organic chemistry, surface science, semiconductor physics, microfabrication, molecular biology etc. The books deals with all the aspects of nanotechnology and its fundamentals and applications, i.e. Nanotechnology, tools and techniques, nanomaterials, nanocomputers, nanocomposites, risks related with the use of nanotechnology and its relation with the environment. Applications of nanotechnology in textiles, solar power, electronics, space explorations and in communication find a reasonable place in the book. This book explains scientific foundations governing the functionality of nanostructures and makes the reader familiar with several basic and application aspects of nano-structured systems. Areas of physics such as nanoelectronics, nanomechanics, nanophotonics and nanoionics have evolved during the last few decades to provide a basic scientific foundation of nanotechnology.

The new millenium has seen the beginnings of the applications of nanotechnology in commercial products, although most applications are limited to the bulk use of passive nanomaterials. Not only will nanotechnology dominate research for years to come, it promises to produce some of the most extraordinary advances in all of applied science. This groundbreaking book provides the first unified treatment of the research, technology, and applications that are fuelling the rapid growth of nanotechnology today. It brings together the physics, quantum mechanics, chemistry, biology, and electronics now blazing new trails - both in the development of nanodevices and in our understanding of physical phenomena never before seen at larger scale. Nanotechnology is helping to considerably improve, even revolutionize, many technology and industry sectors: information technology, energy, environmental science, medicine, homeland security, food safety, and transportation, among many others. Most benefits of nanotechnology depend on the fact that it is possible to tailor the essential structures of materials at the nanoscale to achieve specific properties, thus greatly extending the well-used toolkits of materials science. Examples include titanium dioxide and zinc oxide nanoparticles in sunscreen, cosmetics and some food products; silver nanoparticles in food packaging, clothing, disinfectants and household appliances such as Silver Nano; carbon nanotubes for stain-resistant textiles; and cerium oxide as a fuel catalyst. Nanotechnology is being used in developing countries to help treat disease and prevent health issues. Using nanotechnology, materials can effectively be made to be stronger, lighter, more durable, more reactive, more sieve-like, or better electrical conductors, among many other traits. There already exist over 800 everyday commercial products that rely on nanoscale materials and processes. In this volume, present and future developments are addressed, as well as a larger number of new industrial and research opportunities deriving from this domain.

This book integrates the overview of knowledge from nanofabrication, nanodevices, nanomechanics, nanotribology, materials science, and reliability engineering in one volume. It will analyse fundamental size-dependent phenomena in several fields of chemistry and physics, which form part of leading research efforts in nanotechnology. The products of advanced nanotechnology that will become available in coming decades promise even more revolutionary applications than the products of current and near-term nanotechnology and they are referred to in this volume. It is suitable for scholars in related fields who require an overview of the fundamental topics in nanotechnology. The main objective of this book is providing an introduction to nanotechnology, such that researchers in other areas too can appreciate this science at a new level and technical administrators can have a bird's eye view of this subject. This book on nanoscience and nanotechnology gathers information about the various tools, methods and systems to provide students, researchers and everyone else an open-source handbook and overview guide to this vast interdisciplinary and expanding field.

Dr. Rita Khare

1

Introduction

Nanotechnology (sometimes shortened to "nanotech") is the manipulation of matter on an atomic, molecular, and supramolecular scale. The earliest, widespread description of nanotechnology referred to the particular technological goal of precisely manipulating atoms and molecules for fabrication of macroscale products, also now referred to as molecular nanotechnology. A more generalized description of nanotechnology was subsequently established by the National Nanotechnology Initiative, which defines nanotechnology as the manipulation of matter with at least one dimension sized from 1 to 100 nanometres. This definition reflects the fact that quantum mechanical effects are important at this quantum-realm scale, and so the definition shifted from a particular technological goal to a research category inclusive of all types of research and technologies that deal with the special properties of matter that occur below the given size threshold. It is therefore common to see the plural form "nanotechnologies" as well as "nanoscale technologies" to refer to the broad range of research and applications whose common trait is size. Because of the variety of potential applications (including industrial and military), governments have invested billions of dollars in nanotechnology research. Through its National Nanotechnology Initiative, the USA has invested 3.7 billion dollars. The European Union has invested 1.2 billion and Japan 750 million dollars.

Nanotechnology as defined by size is naturally very broad, including fields of science as diverse as surface science, organic chemistry, molecular biology, semiconductor physics, microfabrication, etc. The associated research and applications are equally diverse, ranging from extensions of conventional device physics to completely new approaches based upon molecular self-assembly, from developing new materials with dimensions on the nanoscale to direct control of matter on the atomic scale.

Scientists currently debate the future implications of nanotechnology. Nanotechnology may be able to create many new materials and devices with a vast range of applications, such as in medicine, electronics, biomaterials and energy production. On the other hand, nanotechnology raises many of the same issues as any new technology, including concerns about the toxicity and environmental impact of nanomaterials, and their potential effects on global economics, as well as speculation about various doomsday scenarios. These concerns have led to a debate among advocacy groups and governments on whether special regulation of nanotechnology is warranted.

Origins

The concepts that seeded nanotechnology were first discussed in 1959 by renowned physicist Richard Feynman in his talk *There's Plenty of Room at the Bottom*, in which he described the possibility of synthesis via direct manipulation of atoms. The term "nano-technology" was first used by Norio Taniguchi in 1974, though it was not widely known.

Inspired by Feynman's concepts, K. Eric Drexler independently used the term "nanotechnology" in his 1986 book *Engines of Creation: The Coming Era of Nanotechnology*, which proposed the idea of a nanoscale "assembler" which would be able to build a copy of itself and of other items of arbitrary complexity with atomic control. Also in 1986, Drexler co-founded The Foresight Institute (with which he is no longer affiliated) to help increase public awareness and understanding of nanotechnology concepts and implications.

Thus, emergence of nanotechnology as a field in the 1980s occurred through convergence of Drexler's theoretical and public work, which developed and popularized a conceptual framework for nanotechnology, and high-visibility experimental advances that drew additional wide-scale attention to the prospects of atomic control of matter.

For example, the invention of the scanning tunneling microscope in 1981 provided unprecedented visualization of individual atoms and bonds, and was successfully used to manipulate individual atoms in 1989. The microscope's developers Gerd Binnig and Heinrich Rohrer at IBM Zurich Research Laboratory received a Nobel Prize in Physics in 1986. Binnig, Quate and Gerber also invented the analogous atomic force microscope that year.

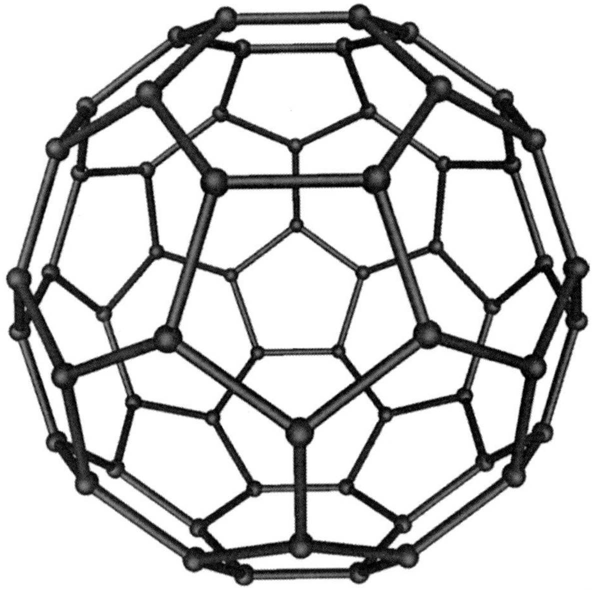

Figure: *Buckminsterfullerene* C_{60}, *also known as the buckyball, is a representative member of the carbon structures known as fullerenes. Members of the fullerene family are a major subject of research falling under the nanotechnology umbrella.*

Fullerenes were discovered in 1985 by Harry Kroto, Richard Smalley, and Robert Curl, who together won the 1996 Nobel Prize in Chemistry. C_{60} was not initially described as nanotechnology; the term was used regarding subsequent work with related graphene tubes (called carbon nanotubes and sometimes called Bucky tubes) which suggested potential applications for nanoscale electronics and devices.

In the early 2000s, the field garnered increased scientific, political, and commercial attention that led to both controversy and progress. Controversies emerged regarding the definitions and potential implications of nanotechnologies, exemplified by the Royal Society's report on nanotechnology. Challenges were raised regarding the feasibility of applications envisioned by advocates of molecular nanotechnology, which culminated in a public debate between Drexler and Smalley in 2001 and 2003.

Meanwhile, commercialization of products based on advancements in nanoscale technologies began emerging. These products are limited to bulk applications of nanomaterials and do not involve atomic control of matter. Some examples include the Silver Nano platform for using silver nanoparticles as an antibacterial agent, nanoparticle-based transparent sunscreens, and carbon nanotubes for stain-resistant textiles. Governments moved to promote and fund research into nanotechnology, beginning in the U.S. with the National Nanotechnology Initiative, which formalized a size-based definition of nanotechnology and established funding for research on the nanoscale.

By the mid-2000s new and serious scientific attention began to flourish. Projects emerged to produce nanotechnology roadmaps which centre on atomically precise manipulation of matter and discuss existing and projected capabilities, goals, and applications.

Fundamental Concepts

Nanotechnology is the engineering of functional systems at the molecular scale. This covers both current work and concepts that are more advanced. In its original sense, nanotechnology refers to the projected ability to construct items from the bottom up, using techniques and tools being developed today to make complete, high performance products.

One nanometre (nm) is one billionth, or 10^{-9}, of a metre. By comparison, typical carbon-carbon bond lengths, or the spacing between these atoms in a molecule, are in the range 0.12–0.15 nm, and a DNA double-helix has a diameter around 2 nm. On the other hand, the smallest cellular life-forms, the bacteria of the genus Mycoplasma, are around 200 nm in length. By convention, nanotechnology is taken as the scale range 1 to 100 nm following the definition used by the National Nanotechnology Initiative in the US. The lower limit is set by the size of atoms (hydrogen has the smallest atoms, which are approximately a quarter of a nm diameter) since nanotechnology must build its devices from atoms and molecules. The upper limit is more or less arbitrary but is around the size that phenomena not observed in larger structures start to become apparent and can be made use of in the nano device. These new phenomena make nanotechnology distinct from devices which are merely miniaturised versions of an equivalent macroscopic device; such devices are on a larger scale and come under the description of microtechnology.

To put that scale in another context, the comparative size of a nanometre to a metre is the same as that of a marble to the size of the earth. Or another way of putting it: a nanometre is the amount an average man's beard grows in the time it takes him to raise the razor to his face.

Two main approaches are used in nanotechnology. In the "bottom-up" approach, materials and devices are built from molecular components which assemble themselves chemically by principles of molecular recognition. In the "top-down" approach, nano-objects are constructed from larger entities without atomic-level control.

Areas of physics such as nanoelectronics, nanomechanics, nanophotonics and nanoionics have evolved during the last few decades to provide a basic scientific foundation of nanotechnology.

Larger to Smaller: a Materials Perspective

Figure: *Image of reconstruction on a clean Gold(100) surface, as visualized using scanning tunneling microscopy. The positions of the individual atoms composing the surface are visible.*

Several phenomena become pronounced as the size of the system decreases. These include statistical mechanical effects, as well as quantum mechanical effects, for example the "quantum size effect" where the electronic properties of solids are altered with great reductions in particle size. This effect does not come into play by going from macro to micro dimensions. However, quantum effects can become significant when the nanometre size range is reached, typically at distances of 100 nanometres or less, the so-called quantum realm. Additionally, a number of physical (mechanical, electrical, optical, etc.) properties change when compared to macroscopic systems. One example is the increase in surface area to volume ratio altering mechanical, thermal and catalytic properties of materials. Diffusion and reactions at nanoscale, nanostructures materials and nanodevices with fast ion transport are generally referred to nanoionics. *Mechanical* properties of nanosystems are of interest in the nanomechanics research. The catalytic activity of nanomaterials also opens potential risks in their interaction with biomaterials.

Materials reduced to the nanoscale can show different properties compared to what they exhibit on a macroscale, enabling unique applications. For instance, opaque substances can become transparent (copper); stable materials can turn combustible (aluminum); insoluble materials may become soluble (gold). A material such as gold, which is chemically inert at normal scales, can serve as a potent chemical catalyst at nanoscales. Much of the fascination with nanotechnology stems from these quantum and surface phenomena that matter exhibits at the nanoscale.

Simple to Complex: a Molecular Perspective

Modern synthetic chemistry has reached the point where it is possible to prepare small molecules to almost any structure. These methods are used today to manufacture a wide variety of useful chemicals such as pharmaceuticals or commercial polymers. This ability raises the question of extending this kind of control to the next-larger level, seeking methods to assemble these single molecules into supramolecular assemblies consisting of many molecules arranged in a well defined manner.

These approaches utilise the concepts of molecular self-assembly and/or supramolecular chemistry to automatically arrange themselves into some useful conformation through a bottom-up approach. The concept of molecular recognition is especially important: molecules can be designed so that a specific configuration or arrangement is favoured due to non-covalent intermolecular forces. The Watson–Crick basepairing rules are a direct result of this, as is the specificity of an enzyme being targeted to a single substrate, or the specific folding of the protein itself. Thus, two or more components can be designed to be complementary and mutually attractive so that they make a more complex and useful whole.

Such bottom-up approaches should be capable of producing devices in parallel and be much cheaper than top-down methods, but could potentially be overwhelmed as the size and complexity of the desired assembly increases. Most useful structures require complex and thermodynamically unlikely arrangements of atoms. Nevertheless, there are many examples of self-assembly based on molecular recognition in biology, most notably Watson–Crick basepairing and enzyme-substrate interactions. The challenge for nanotechnology is whether these principles can be used to engineer new constructs in addition to natural ones.

Molecular Nanotechnology: a Long-Term View

Molecular nanotechnology, sometimes called molecular manufacturing, describes engineered nanosystems (nanoscale machines) operating on the molecular scale. Molecular nanotechnology is especially associated with the molecular assembler, a machine that can produce a desired structure or device atom-by-atom using the principles of mechanosynthesis. Manufacturing in the context of productive nanosystems is not related to, and should be clearly distinguished from, the conventional technologies used to manufacture nanomaterials such as carbon nanotubes and nanoparticles.

When the term "nanotechnology" was independently coined and popularized by Eric Drexler (who at the time was unaware of an earlier usage by Norio Taniguchi) it referred to a future manufacturing technology based on molecular machine systems. The premise was that molecular scale biological analogies of traditional machine components demonstrated molecular machines were possible: by the countless examples found in biology, it is known that sophisticated, stochastically optimised biological machines can be produced.

It is hoped that developments in nanotechnology will make possible their construction by some other means, perhaps using biomimetic principles. However, Drexler and other researchers have proposed that advanced nanotechnology, although perhaps initially implemented by biomimetic means, ultimately could be based on mechanical engineering principles, namely, a manufacturing technology based on the mechanical functionality of these components (such as gears, bearings, motors, and structural members) that would enable programmable, positional assembly to atomic specification. The physics and engineering performance of exemplar designs were analyzed in Drexler's book *Nanosystems*.

In general it is very difficult to assemble devices on the atomic scale, as one has to position atoms on other atoms of comparable size and stickiness. Another view, put forth by Carlo Montemagno, is that future nanosystems will be hybrids of silicon technology and biological molecular machines. Richard Smalley argued that mechanosynthesis are impossible due to the difficulties in mechanically manipulating individual molecules.

This led to an exchange of letters in the ACS publication Chemical & Engineering News in 2003. Though biology clearly demonstrates that molecular machine systems are possible, non-biological molecular machines are today only in their infancy. Leaders in research on non-biological molecular machines are Dr. Alex Zettl and his colleagues at Lawrence Berkeley Laboratories and UC Berkeley. They have constructed at least three distinct molecular devices whose motion is controlled from the desktop with changing voltage: a nanotube nanomotor, a molecular actuator, and a nanoelectromechanical relaxation oscillator.

An experiment indicating that positional molecular assembly is possible was performed by Ho and Lee at Cornell University in 1999. They used a scanning tunneling microscope to move an individual carbon monoxide molecule (CO) to an individual iron atom (Fe) sitting on a flat silver crystal, and chemically bound the CO to the Fe by applying a voltage.

Current Research

Nanomaterials

The nanomaterials field includes subfields which develop or study materials having unique properties arising from their nanoscale dimensions.

- Interface and colloid science has given rise to many materials which may be useful in nanotechnology, such as carbon nanotubes and other fullerenes, and various nanoparticles and nanorods. Nanomaterials with fast ion transport are related also to nanoionics and nanoelectronics.
- Nanoscale materials can also be used for bulk applications; most present commercial applications of nanotechnology are of this flavour.
- Progress has been made in using these materials for medical applications.
- Nanoscale materials such as nanopillars are sometimes used in solar cells which combats the cost of traditional Silicon solar cells.
- Development of applications incorporating semiconductor nanoparticles to be used in the next generation of products, such as display technology, lighting, solar cells and biological imaging.

Bottom-up Approaches

These seek to arrange smaller components into more complex assemblies.

- DNA nanotechnology utilises the specificity of Watson–Crick basepairing to construct well-defined structures out of DNA and other nucleic acids.
- Approaches from the field of "classical" chemical synthesis (inorganic and organic synthesis) also aim at designing molecules with well-defined shape (e.g. bis-peptides).
- More generally, molecular self-assembly seeks to use concepts of supramolecular chemistry, and molecular recognition in

particular, to cause single-molecule components to automatically arrange themselves into some useful conformation.

- Atomic force microscope tips can be used as a nanoscale "write head" to deposit a chemical upon a surface in a desired pattern in a process called dip pen nanolithography.

  This technique fits into the larger subfield of nanolithography.

Top-down Approaches

These seek to create smaller devices by using larger ones to direct their assembly.

- Many technologies that descended from conventional solid-state silicon methods for fabricating microprocessors are now capable of creating features smaller than 100 nm, falling under the definition of nanotechnology.

  Giant magnetoresistance-based hard drives already on the market fit this description, as do atomic layer deposition (ALD) techniques. Peter Grünberg and Albert Fert received the Nobel Prize in Physics in 2007 for their discovery of Giant magnetoresistance and contributions to the field of spintronics.
- Solid-state techniques can also be used to create devices known as nanoelectromechanical systems or NEMS, which are related to microelectromechanical systems or MEMS.
- Focused ion beams can directly remove material, or even deposit material when suitable precursor gasses are applied at the same time. For example, this technique is used routinely to create sub-100 nm sections of material for analysis in Transmission electron microscopy.
- Atomic force microscope tips can be used as a nanoscale "write head" to deposit a resist, which is then followed by an etching process to remove material in a top-down method.

Functional Approaches

These seek to develop components of a desired functionality without regard to how they might be assembled.

- Molecular scale electronics seeks to develop molecules with useful electronic properties. These could then be used as single-molecule components in a nanoelectronic device.
- Synthetic chemical methods can also be used to create synthetic molecular motors, such as in a so-called nanocar.

Biomimetic Approaches

- Bionics or biomimicry seeks to apply biological methods and systems found in nature, to the study and design of engineering systems and modern technology. Biomineralization is one example of the systems studied.
- Bionanotechnology is the use of biomolecules for applications in nanotechnology, including use of viruses and lipid assemblies. Nanocellulose is a potential bulk-scale application.

Speculative

These subfields seek to anticipate what inventions nanotechnology might yield, or attempt to propose an agenda along which inquiry might progress. These often take a big-picture view of nanotechnology, with more emphasis on its societal implications than the details of how such inventions could actually be created.

- Molecular nanotechnology is a proposed approach which involves manipulating single molecules in finely controlled, deterministic ways. This is more theoretical than the other subfields, and many of its proposed techniques are beyond current capabilities.
- Nanorobotics centres on self-sufficient machines of some functionality operating at the nanoscale. There are hopes for applying nanorobots in medicine, but it may not be easy to do such a thing because of several drawbacks of such devices. Nevertheless, progress on innovative materials and methodologies has been demonstrated with some patents granted about new nanomanufacturing devices for future commercial applications, which also progressively helps in the development towards nanorobots with the use of embedded nanobioelectronics concepts.
- Productive nanosystems are "systems of nanosystems" which will be complex nanosystems that produce atomically precise parts for other nanosystems, not necessarily using novel nanoscale-emergent properties, but well-understood fundamentals of manufacturing. Because of the discrete (i.e. atomic) nature of matter and the possibility of exponential growth, this stage is seen as the basis of another industrial revolution. Mihail Roco, one of the architects of the USA's National Nanotechnology Initiative, has proposed four states of nanotechnology that seem to parallel the technical progress of the Industrial Revolution, progressing from passive nanostructures to active nanodevices to complex nanomachines and ultimately to productive nanosystems.

- Programmable matter seeks to design materials whose properties can be easily, reversibly and externally controlled though a fusion of information science and materials science.
- Due to the popularity and media exposure of the term nanotechnology, the words picotechnology and femtotechnology have been coined in analogy to it, although these are only used rarely and informally.

Tools and Techniques

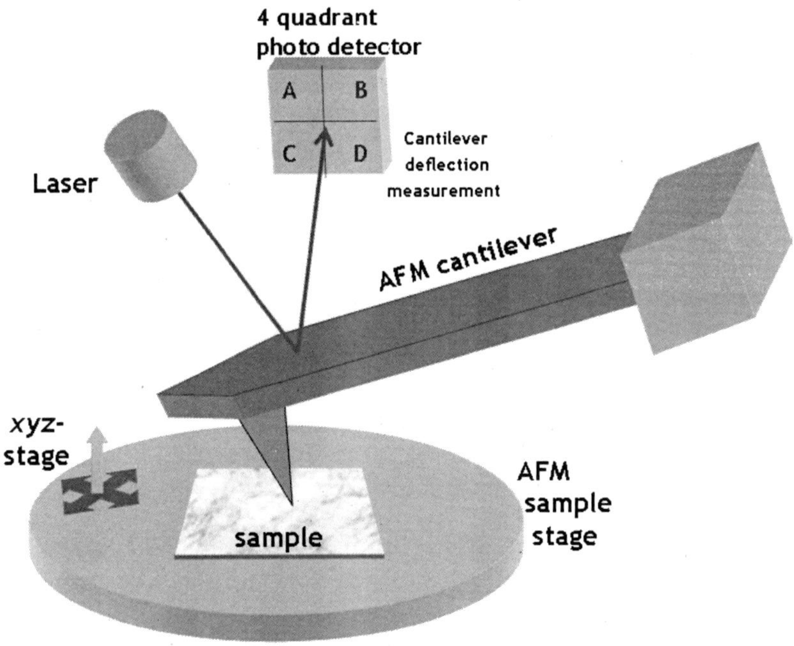

Figure: *Typical AFM setup. A microfabricated cantilever with a sharp tip is deflected by features on a sample surface, much like in a phonograph but on a much smaller scale. A laser beam reflects off the backside of the cantilever into a set of photodetectors, allowing the deflection to be measured and assembled into an image of the surface.*

There are several important modern developments. The atomic force microscope (AFM) and the Scanning Tunneling Microscope (STM) are two early versions of scanning probes that launched nanotechnology. There are other types of scanning probe microscopy. Although conceptually similar to the scanning confocal microscope developed by Marvin Minsky in 1961 and the scanning acoustic microscope (SAM) developed by Calvin Quate and coworkers in the 1970s, newer scanning probe microscopes have much higher resolution, since they are not limited by the wavelength of sound or light. The tip of a scanning probe can also be used to manipulate nanostructures (a process called positional assembly). Feature-oriented scanning methodology suggested by Rostislav Lapshin appears to be a promising way to implement these nanomanipulations in automatic mode. However, this is still a slow process because of low scanning velocity of the microscope.

Various techniques of nanolithography such as optical lithography, X-ray lithography dip pen nanolithography, electron beam lithography or nanoimprint lithography were also developed. Lithography is a top-down fabrication technique where a bulk material is reduced in size to nanoscale pattern.

Another group of nanotechnological techniques include those used for fabrication of nanotubes and nanowires, those used in semiconductor fabrication such as deep ultraviolet lithography, electron beam lithography, focused ion beam machining, nanoimprint lithography, atomic layer deposition, and molecular vapor deposition, and further including molecular self-assembly techniques such as those employing di-block copolymers. The precursors of these techniques preceded the nanotech era, and are extensions in the development of scientific advancements rather than techniques which were devised with the sole purpose of creating nanotechnology and which were results of nanotechnology research.

The top-down approach anticipates nanodevices that must be built piece by piece in stages, much as manufactured items are made. Scanning probe microscopy is an important technique both for characterization and synthesis of nanomaterials. Atomic force microscopes and scanning tunneling microscopes can be used to look at surfaces and to move atoms around. By designing different tips for these microscopes, they can be used for carving out structures on surfaces and to help guide self-assembling structures. By using, for example, feature-oriented scanning approach, atoms or molecules can be moved around on a surface with scanning probe microscopy techniques. At present, it is expensive and time-consuming for mass production but very suitable for laboratory experimentation.

In contrast, bottom-up techniques build or grow larger structures atom by atom or molecule by molecule. These techniques include chemical synthesis, self-assembly and positional assembly. Dual polarisation interferometry is one tool suitable for characterisation of self assembled thin films. Another variation of the bottom-up approach is molecular beam epitaxy or MBE. Researchers at Bell Telephone Laboratories like John R. Arthur. Alfred Y. Cho, and Art C. Gossard developed and implemented MBE as a research tool in the late 1960s and 1970s. Samples made by MBE were key to the discovery of the fractional quantum Hall effect for which the 1998 Nobel Prize in Physics was awarded. MBE allows scientists to lay down atomically precise layers of atoms and, in the process, build up complex structures. Important for research on semiconductors, MBE is also widely used to make samples and devices for the newly emerging field of spintronics.

However, new therapeutic products, based on responsive nanomaterials, such as the ultradeformable, stress-sensitive Transfersome vesicles, are under development and already approved for human use in some countries.

Applications

As of August 21, 2008, the Project on Emerging Nanotechnologies estimates that over 800 manufacturer-identified nanotech products are publicly available, with new ones hitting the market at a pace of 3–4 per week. The project lists all of the products in a publicly accessible online database. Most applications are limited to the use of "first generation" passive nanomaterials which includes titanium dioxide in sunscreen, cosmetics, surface coatings, and some food products; Carbon allotropes used to produce gecko tape; silver in food packaging, clothing, disinfectants and household appliances; zinc oxide in sunscreens and cosmetics, surface coatings, paints and outdoor furniture varnishes; and cerium oxide as a fuel catalyst.

Further applications allow tennis balls to last longer, golf balls to fly straighter, and even bowling balls to become more durable and have a harder surface. Trousers and socks have been infused with nanotechnology so that they will last longer and keep people cool in the summer. Bandages are being infused with silver nanoparticles to heal cuts faster. Cars are being manufactured with nanomaterials so they may need fewer metals and less fuel to operate in the future. Video game consoles and personal computers may become cheaper, faster, and contain more memory thanks to nanotechnology. Nanotechnology may have the ability to make existing medical applications cheaper and easier to use in places like the general practitioner's office and at home.

The National Science Foundation (a major distributor for nanotechnology research in the United States) funded researcher David Berube to study the field of nanotechnology. His findings are published in the monograph Nano-Hype: The Truth Behind the Nanotechnology Buzz. This study concludes that much of what is sold as "nanotechnology" is in fact a recasting of straightforward materials science, which is leading to a "nanotech industry built solely on selling nanotubes, nanowires, and the like" which will "end up with a few suppliers selling low margin products in huge volumes." Further applications which require actual manipulation or arrangement of nanoscale components await further research. Though technologies branded with the term 'nano' are sometimes little related to and fall far short of the most ambitious and transformative technological goals of the sort in molecular manufacturing proposals, the term still connotes such ideas. According to Berube, there may be a danger that a "nano bubble" will form, or is forming already, from the use of the term by scientists and entrepreneurs to garner funding, regardless of interest in the transformative possibilities of more ambitious and far-sighted work.

Implications

An area of concern is the effect that industrial-scale manufacturing and use of nanomaterials would have on human health and the environment, as suggested by nanotoxicology research. For these reasons, some groups advocate that nanotechnology be regulated by governments. Others counter that overregulation would stifle scientific research and the development of beneficial innovations. Public health research agencies, such as the National Institute for Occupational Safety and Health are actively conducting research on potential health effects stemming from exposures to nanoparticles.

Some nanoparticle products may have unintended consequences. Researchers have discovered that bacteriostatic silver nanoparticles used in socks to reduce foot odor are being released in the wash. These particles are then flushed into the waste water stream and may destroy bacteria which are critical components of natural ecosystems, farms, and waste treatment processes.

Public deliberations on risk perception in the US and UK carried out by the Centre for Nanotechnology in Society found that participants were more positive about nanotechnologies for energy applications than for health applications, with health applications raising moral and ethical dilemmas such as cost and availability.

Experts, including director of the Woodrow Wilson Centre's Project on Emerging Nanotechnologies David Rejeski, have testified that successful commercialization depends on adequate oversight, risk research strategy, and public engagement. Berkeley, California is currently the only city in the United States to regulate nanotechnology; Cambridge, Massachusetts in 2008 considered enacting a similar law, but ultimately rejected it. Relevant for both research on and application of nanotechnologies, the insurability of nanotechnology is contested. Without state regulation of nanotechnology, the availability of private insurance for potential damages is seen as necessary to ensure that burdens are not socialised implicitly.

Health and Environmental Concerns

Nanofibres are used in several areas and in different products, in everything from aircraft wings to tennis rackets. Inhaling airborne nanoparticles and nanofibres may lead to a number of pulmonary diseases, e.g. fibrosis. Researchers have found that when rats breathed in nanoparticles, the particles settled in the brain and lungs, which led to significant increases in biomarkers for inflammation and stress response and that nanoparticles induce skin aging through oxidative stress in hairless mice.

A two-year study at UCLA's School of Public Health found lab mice consuming nano-titanium dioxide showed DNA and chromosome damage to a degree "linked to all the big killers of man, namely cancer, heart disease, neurological disease and aging".

A major study published more recently in Nature Nanotechnology suggests some forms of carbon nanotubes – a poster child for the "nanotechnology revolution" – could be as harmful as asbestos if inhaled in sufficient quantities. Anthony Seaton of the Institute of Occupational Medicine in Edinburgh, Scotland, who contributed to the article on carbon nanotubes said "We know that some of them probably have the potential to cause mesothelioma. So those sorts of materials need to be handled very carefully." In the absence of specific regulation forthcoming from governments, Paull and Lyons (2008) have called for an exclusion of engineered nanoparticles in food. A newspaper article reports that workers in a paint factory developed serious lung disease and nanoparticles were found in their lungs.

Regulation

Calls for tighter regulation of nanotechnology have occurred alongside a growing debate related to the human health and safety risks of nanotechnology. There is significant debate about who is responsible for the regulation of nanotechnology. Some regulatory agencies currently cover some nanotechnology products and processes (to varying degrees) – by "bolting on" nanotechnology to existing regulations – there are clear gaps in these regimes. Davies (2008) has proposed a regulatory road map describing steps to deal with these shortcomings.

Stakeholders concerned by the lack of a regulatory framework to assess and control risks associated with the release of nanoparticles and nanotubes have drawn parallels with bovine spongiform encephalopathy ("mad cow" disease), thalidomide, genetically modified food, nuclear energy, reproductive technologies, biotechnology, and asbestosis. Dr. Andrew Maynard, chief science advisor to the Woodrow

Wilson Centre's Project on Emerging Nanotechnologies, concludes that there is insufficient funding for human health and safety research, and as a result there is currently limited understanding of the human health and safety risks associated with nanotechnology. As a result, some academics have called for stricter application of the precautionary principle, with delayed marketing approval, enhanced labelling and additional safety data development requirements in relation to certain forms of nanotechnology.

The Royal Society report identified a risk of nanoparticles or nanotubes being released during disposal, destruction and recycling, and recommended that "manufacturers of products that fall under extended producer responsibility regimes such as end-of-life regulations publish procedures outlining how these materials will be managed to minimize possible human and environmental exposure" (p. xiii). Reflecting the challenges for ensuring responsible life cycle regulation, the Institute for Food and Agricultural Standards has proposed that standards for nanotechnology research and development should be integrated across consumer, worker and environmental standards. They also propose that NGOs and other citizen groups play a meaningful role in the development of these standards.

The Centre for Nanotechnology in Society has found that people respond to nanotechnologies differently, depending on application – with participants in public deliberations more positive about nanotechnologies for energy than health applications – suggesting that any public calls for nano regulations may differ by technology sector.

Fail-Safes in Nanotechnology

Fail-safes in nanotechnology are devices or features integrated with nanotechnology which, in the event of failure, respond in a way that will cause no harm, or at least a minimum of harm, to other devices or personnel. Fail-safe principles are governed by national standards and engineering practices, and are widely used in conventional engineering design. It is possible to scale down macro-scale fail-safe principles and devices for similar applications at the nano-scale. The use of fail-safes in nanotechnology applications supports social acceptance of those applications by reducing the risks to users; as of 2009, there are both theoretical and practical ways to implement fail-safe designs in nanotechnology.

A predominant challenge to the social acceptance of nanotechnology is concerned with the medical use of nanostructures in the human body. While any structure for medical use would be developed to be biocompatible and harmless, sound engineering design must take into account all possibilities of failure. Thus, the design would include ways to manipulate the structures in the body in the event of failure.

Ferrous Nanoparticles

Many researchers are looking into creating nano-scale robots ("nanobots"), for the purpose of undertaking tasks where only robots on the nano scale can be used, such as inside the human body. These robots would have the ability to construct other nanostructures or perform medical procedures, and will be introduced into the body via an injection. The robots' shells and circuits would be made of ferrous nanoparticles so that a magnetic field could be used to prevent or manipulate their movement. In case of failure or malfunction, a small EMP or an MRI could be used to deactivate the nanobots. Both techniques induce an electromagnetic field, corrupting the memory and shorting out the circuitry of any electronic device within range.

Amino Acids

Researchers are pursuing the building of nanostructures using amino acids. Nanostructures that are created using amino acids are constructed using only synthetic types of amino acids, which tags these structures with unique molecules. These engineered amino acids essentially form synthetic proteins that differ from the naturally occurring proteins in the human body. This difference in the engineered amino acids makes these proteins easy to isolate and target. In case of failure or malfunction, it is possible to identify these proteins using the specifically targeted molecules, which act as a flag to indicate the location of the target. Then, another mechanism would be used to isolate them and deactivate them.

DNA

DNA within our bodies naturally breaks down, replicates itself, and rebuilds itself every time a cell divides. These processes are all controlled and completed by various enzymes. DNA molecules are composed of corresponding base pair nucleotides in a double-helix formation, which makes these processes very efficient, accurate, and predictable. Due to the ease with which DNA molecules can be fashioned, many publications in the academic society are geared towards creating nanostructures using DNA. With a DNA-based nano-device, synthetic proteins could be created, designed to deactivate a nano-device. These synthetic proteins would be injected into the body to break down the DNA and render a nano-device harmless in the event of a malfunction.

Biological proteins within the human body serve three main functions: they are structural building blocks, enzymes, and facilitate cellular signalling. Synthetic proteins could be developed as a form of indicator and attached to a DNA-based nano-device. This indicator would then be used for the purpose of monitoring nano-devices in the human body. If all DNA-based nano-devices were closely monitored in the human body, they could be controlled quickly in the event of a malfunction.

Programming

In nanotechnology, particularly in nanobots, the need for a sound programming architecture is very important due to a potentially higher risk of damage in the event of a malfunction. A two-layer approach can be used to control nano-devices: (1) by providing a preprogrammed fail-safe functionality in case of anticipated failures; and (2) a remote-controlled override for use in unforeseen situations. The "remote"-controlled nano-device would require a specialist in the room, to guide the nanobot throughout the procedure.

Cellular Engineering

Many researchers are developing methods that use bacteria to deliver drugs. These bacteria can be "programmed" to perform a specific task, and can be directed to go to targeted locations in the body. However, the bacteria may damage healthy organs or fail to deliver the medicine to the sick organ in the case of a malfunction. In such cases, a fail-safe mechanism is required to neutralize the bacteria and prevent damage. An antibiotic is generally suitable as the fail-safe agent.

Nanotechnology Environment and Energy

The environmental impact of nanotechnology is the possible effects that the use of nanotechnological materials and devices will have on the environment. As nanotechnology is an emerging field, there is great debate regarding to what extent industrial and commercial use of nanomaterials will affect organisms and ecosystems. Nanotechnology's environmental impact can be split into two aspects: the potential for nanotechnological innovations to help improve the environment, and the possibly novel type of pollution that nanotechnological materials might cause if released into the environment.

Nanopollution

Nanopollution is a generic name for waste generated by nanodevices or during the nanomaterials manufacturing process. Ecotoxicological impacts of nanoparticles and the potential for bioaccumulation in plants and microorganisms is a subject of current research, as nanoparticles are considered to present novel environmental impacts. Of the US$710 million spent in 2002 by the U.S. government on nanotechnology research, $500,000 was spent on environmental impact assessments.

Figure: *Groups opposing the installation of nanotechnology laboratories in Grenoble, France, have spraypainted their opposition on a former fortress above the city*

The capacity for nanoparticles to function as a transport mechanism also raises concern about the transport of heavy metals and other environmental contaminants. Two areas of concern can be identified. First, in their free form nanoparticles can be released into the air or water during production, or production accidents, or as waste by-product of production, and ultimately accumulate in the soil, water, or plant life. Second, in fixed form, where they are part of a manufactured substance or product, they will ultimately have to be recycled or disposed of as waste.

Scrinis raises concerns about nano-pollution, and argues that it is not currently possible to "precisely predict or control the ecological impacts of the release of these nano-products into the environment." A May 2007 Report to the UK Department for Environment, Food and Rural Affairs noted concerns about the toxicological impacts of nanoparticles in relation to both hazard and exposure. The report recommended comprehensive toxicological testing and independent performance tests of fuel additives. Risks have been identified by

Uskokovic in 2007. Concerns have also been raised about Silver Nano technology used by Samsung in a range of appliances such as washing machines and air purifiers.

Life Cycle Responsibility

To properly assess the health hazards of engineered nanoparticles the whole life cycle of these particles needs to be evaluated, including their fabrication, storage and distribution, application and potential abuse, and disposal. The impact on humans or the environment may vary at different stages of the life cycle.

The Royal Society report identified a risk of nanoparticles or nanotubes being released during disposal, destruction and recycling, and recommended that "manufacturers of products that fall under extended producer responsibility regimes such as end-of-life regulations publish procedures outlining how these materials will be managed to minimize possible human and environmental exposure" (p.xiii). Reflecting the challenges for ensuring responsible life cycle regulation, the Institute for Food and Agricultural Standards has proposed standards for nanotechnology research and development should be integrated across consumer, worker and environmental standards. They also propose that NGOs and other citizen groups play a meaningful role in the development of these standards.

Environmental Benefits of Nanotechnology

Energy: Nanotechnology could potentially have a great impact on clean energy production. Research is underway to use nanomaterials for purposes including more efficient solar cells, practical fuel cells, and environmentally friendly batteries. The most advanced nanotechnology projects related to energy are: storage, conversion, manufacturing improvements by reducing materials and process rates, energy saving (by better thermal insulation for example), and enhanced renewable energy sources.

Current commercially available solar cells have low efficiencies of 15-20%. Research is ongoing to use nanowires and other nanostructured materials with the hope of to create cheaper and more efficient solar cells than are possible with conventional planar silicon solar cells. It is believed that these nanoelectronics-based devices will enable more efficient solar cells, and would have a great effect on satisfying global energy needs.

Another example for an environmentally friendly form of energy is the use of fuel cells powered by hydrogen. Probably the most prominent nanostructured material in fuel cells is the catalyst consisting of carbon supported noble metal particles with diameters of 1-5 nm. Suitable materials for hydrogen storage contain a large number of small nanosized pores. Nanotechnology may also find applications in batteries. Because of the relatively low energy density of conventional batteries the operating time is limited and a replacement or recharging is needed, and the huge number of spent batteries represent a disposal problem. The use of nanomaterials may enable batteries with higher energy content or supercapacitors with a higher rate of recharging, which could be helpful for the battery disposal problem.

Water Filtration and Remediation

A strong influence of nanochemistry on waste-water treatment, air purification and energy storage devices is to be expected. Mechanical or chemical methods can be used for effective filtration techniques. One class of filtration techniques is based on the use of membranes with suitable hole sizes, whereby the liquid is pressed through the membrane. Nanoporous membranes are suitable for a mechanical filtration with extremely small pores smaller than 10 nm ("nanofiltration") and may be composed of nanotubes. Nanofiltration is mainly used for the removal of ions or the separation of different fluids.

Magnetic nanoparticles offer an effective and reliable method to remove heavy metal contaminants from waste water by making use of magnetic separation techniques. Using nanoscale particles increases the efficiency to absorb the contaminants and is comparatively inexpensive compared to traditional precipitation and filtration methods.

Some water-treatment devices incorporating nanotechnology are already on the market, with more in development. Low-cost nanostructured separation membranes methods have been shown to be effective in producing potable water in a recent study. Nanoscale iron particles have also shown potential as a detoxifying agent for cleaning environmental contaminents from brownfield sites.

Solar Power Using Nanotechnology

Using nanoparticles in the manufacture of solar cells has the following benefits:

- Reduced manufacturing costs as a result of using a low temperature process similar to printing instead of the high temperature vacuum deposition process typically used to produce conventional cells made with crystalline semiconductor material.
- Reduced installation costs achieved by producing flexible rolls instead of rigid crystalline panels. Cells made from semiconductor thin films will also have this characteristic.

- Currently available nanotechnology solar cells are not as efficient as traditional ones, however their lower cost offsets this. In the long term nanotechnology versions should both be lower cost and, using quantum dots, should be able to reach higher efficiency levels than conventional ones.

Solar Cells: Nanotechnology Applications under Development

Researchers at Michigan Technological University have developed a honeycomb like structure of graphene in which the graphene sheets are held apart by lithium carbonate. They have used this "3D graphene" to replace the platinum in a dye sensitized solar cell and achieved 7.8 percent conversion of sunlight to electricity. Researchers at Los Alamos National Lab have demonstrated a solar cell that uses a copper indium selenide sulfide quantum dots. Unlike quantum dots containing lead or cadium the copper based quantum dot is non-toxic as well as low cost.

Researchers at MIT are studying solar cells made from single molecule thick sheets of graphene and materials such as molybdenum diselenide. They are predicting that this type of solar cells could produce up to 1000 times as much more power for a given weigh of material than conventional solar cells. They have completed computer modelling and are working on building the solar cells.

Researchers at MIT have developed a solar cell using graphene coated with zinc oxide nanowires. The researchers believe that this method will allow the production of low cost flexible solar cells at high enough efficiency to be competive. Researchers are using a method called Aerotaxy to grow semiconducting nanowires on gold nanoparticles. They plan to use self assembly techniques to align the nanowires on a substrate; forming a solar cell or other electrical devices. The gold nanoparticles replace the silicon substrate on which conventional semiconductor based solar cells are built.

Researchers at Princeton University have developed a solar cell that almost eliminates losses due to reflection of light. They use a 30 nanometre thick gold nanomesh (a layer with a regular pattern of 175 nanometre diameter holes) along with an active layer thinner than the wavelength of light. They found that this combination traps most of the light in the solar cell, increasing the efficiency of the cell.

Researchers at Duke University are developing another method to reduce losses due to the reflection of light. In this method the combination of silver nanocubes scattered over a thin gold layer reduce losses due to reflection. Researchers at the Nies Bohr Institute have determined that sunlight can be concentrated in nanowires due to a resonance effect. This effect can result in more efficient solar cells, allowing more of the energy from the sun to be converted to electricity.

An inexpensive solar cell designed to produce hydrogen is being developed using iron oxide nanoparticles. Combining silver nanowires, titanium dioxide nanoparticles and a polymer that absorbs infrared light to make a solar cell that is about 70 percent transparent to visible light, allowing it to be used in windows.

Titanium dioxide nanotubes filled with a polymer to form low cost solar cells Combining lead selenide quantum dots with titanium dioxide to form higher efficiency solar cells.

Combining carbon nanotubes and buckyballs to produce solar cells. Some researchers combine the nanotubes and buckyballs with a polymer, while another group of researchers are only using nanotubes and buckyballs. A third research group is also using nanotubes and buckballs along with graphene to build a solar cell.

Researchers at Stanford University have found a way to trap light in organic solar cells. The idea is that the longer light is in the solar cell the more electrons will be generated. The researchers found that by making the organic layer much thinner than the wavelength of light and sandwiching the organic layer between a mirror layer and a rough layer the light stayed in the solar cell longer and excited more electrons.

Semiconductor nanoparticles applied in a low temperature printing process that results in low cost solar cells.

Organic molecules to lower costs: Using light absorbing nanowires embedded in a flexible polymer film is another method being developed to produce low cost flexible solar panels. Using light absorbing graphene sheets to produce low cost solar panels

Organic solar cells that are self repairing: Organic solar cells that can be applied by spray painting, possibly turning the surface of a car into a solar cell. Researchers at Lawrence Berkeley have demonstrated an inexpensive process for making solar cells. These solar cells are composed of cadmium sulfide nanowires coated with copper sulfide.

Solar cells that can be installed as a coating on windows or other building materials, referred to as "Building Integrated Photovoltaic's".

Nanotechnology and Textiles

Nanofabrics are textiles engineered with small particles that give ordinary materials advantageous properties such as superhydrophobicity (extreme water resistance), odor and moisture elimination, increased elasticity and strength, and bacterial resistance. Depending on the desired property, a nanofabric is either constructed from nanoscopic fibres called nanofibres, or is formed by applying a solution containing nanoparticles to a regular fabric.

Nanofabrics research is an interdisciplinary effort involving bioengineering, molecular chemistry, physics, electrical engineering, computer science, and systems engineering. Applications of nanofabrics have the potential to revolutionize textile manufacturing and areas of medicine such as drug delivery and tissue engineering.

Introduction

A fibre that has a width of less than 1000 nanometres (1000 nm or 1 μm) is generally defined as a nanofibre. A nanoparticle is defined as a small group of atoms or molecules with a radius of less than 100 nanometres (100 nm). Particles on the nanoscale have a very high surface area to volume ratio, whereas this ratio is much lower for objects on the macroscopic scale.

A high relative surface area means that a large proportion of a particle's mass exists on its surface, so nanofibres and nanoparticles show a greater level of interaction with other materials. The high surface area to volume ratio observed in very small particles is what makes it possible to create many special properties exhibited by nanofabrics.

The use of nanoparticles and nanofibres to produce specialised nanofabrics became a subject of interest after the sol-gel and electrospinning techniques were fully developed in the 1980s. Since 2000, dramatic increases in global funding have accelerated research efforts in nanotechnology, including nanofabrics research.

Sol-gel

The sol-gel process is used to create gel-like solutions which can be applied to textiles as a liquid finish to create nanofabrics with novel properties. The process begins with dissolving nanoparticles in a liquid solvent (often an alcohol). Once dissolved, several chemical reactions take place that cause the nanoparticles to grow and establish a network throughout the liquid.

The network transforms the solution into a colloid (a suspension of solid particles in a liquid) with a gelatinous texture. Finally, the colloid must go through a drying process to remove excess solvent from the mixture before it can be used to treat fabrics. The sol-gel process is used in a similar fashion to make polymer nanofibres, which are long, ultra-thin chains of proteins bonded together.

Electrospinning

Electrospinning extracts nanofibres from polymer solutions (synthesized by the sol-gel process) and collects them to form nonwoven nanofabrics. A strong electric field is applied to the solution to charge the polymer strands. The solution is put into a syringe and aimed at an oppositely charged collector plate.

When the force of attraction between the polymer nanofibres and the collector plate exceed the surface tension of the solution, the nanofibres are released from the solution and deposit onto the collector plate. The deposited fibres form a porous nanofabric that can aid in drug delivery and tissue engineering depending on the type of polymer used.

Applications

Textile Manufacturing: When nanoengineered coatings are applied to fabrics, the nanoparticles readily form bonds with the fibres of the material. The high surface area relative to the volume of particles increases their chemical reactivity, allowing them to stick to materials more permanently. Fabrics treated with nanoparticle coatings during manufacturing produce materials that kill bacteria, eliminate moisture and odor, and prevent static electricity.

Polymer nanofibre coatings applied to textiles bond to the material at one end of the polymer, forming a surface of tiny, hair-like structures. The polymer "hairs" create a thin layer that prevents liquids from making contact with the actual fabric. Nanofabrics with dirt-proof, stain-proof, and superhydrophobic properties are possible as a result of the layer formed by polymer nanofibres.

Development of nanofabrics for use in the clothing and textiles industry is still in its early stages. Some applications such as bacteria-resistant clothing are not yet practical from an economic standpoint. For example, a Cornell University student's prototype for a bactericidal jacket cost $10,000 alone, so it may be a long time before nanofabric clothing is on the market.

Drug Delivery: Nanofabrics used in medicine can deliver antibiotics, anticancer drugs, proteins, and DNA in precise quantities. Electrospinning creates porous nanofabrics that can be loaded with the desired drug which are then applied to the tissue of the targeted area. The drug passes through the tissue by diffusion, a process in which substances move through a membrane from high to low concentration. The rate at which the drug is administered can be changed by altering the composition of the nanofabric.

Tissue Engineering: Nonwoven fabrics made by electrospinning have the potential to assist in the growth of organ tissue, bone, neurons, tendons, and ligaments. Polymer nanofabrics can act either as a scaffold to support damaged tissue or as a synthetic substitute for actual tissue. Depending on the function, the nanofabric can be made of natural or synthetic polymers, or a combination of both.

Environmental Implications: As nanotechnology advances, many studies have been conducted to determine the effects nanoengineered materials can have on the environment. Most textiles can lose up to 20% of their mass during their lifetime, so nanoparticles used in production of nanofabrics are at risk of being released into the air and waterways.

Nano-silver is expected to have as much as 49.5% of its global production taken by the nanotextiles industry due to its antibacterial properties. It is predicted that 20% of the nano-silver used in the nanofabrics industry will be released into waterways which could cause harm to microorganisms. However, more than 90% of nano-silver is removed during treatment at wastewater facilities, so it is likely that the environmental impact will be minimal. A study on aluminum oxide nanoparticles showed that inhalation caused inflammation in rat lungs. Aluminum oxide nanoparticles are not used in large quantity, so its health risks are negligible. Other studies conducted for nanoparticles suggest that their environmental impact should be low as the nanotextiles industry continues to grow.

Aquamelt

An aquamelt is a naturally hydrated polymeric material that is able to solidify at environmental temperatures through a controlled stress input (be it mechanical or chemical).

They are unique in being able to "lock in" work applied to them through an alteration in hydrogen bonding, which enables them to be processed with approximately 1000 times less energy than standard polymers. This has been recently shown for an archetypal biopolymer, silk, however the mechanism for solidification is thought to be inherent to many other biological materials.

Discovery and Mechanism

Aquamelts were defined as a new class of polymeric material as a result of a comparison between the spinning feedstock of the Chinese silkworm (*Bombyx mori*) and molten high-density polyethylene (HDPE) using Shear Induced Polarised Light Imaging (SIPLI).

The current understanding of shear induced fibrillation requires polymer chains to undergo the following series of steps i) long-chain molecules are stretched, ii) and form persistent point nuclei, which iii) align under flow into rows and then iv) grow to create a crystalline fibrils. For these fibrils to remain, the temperature of the sample must be lowered to below the polymers melt point. This process is analogous to the fibrilogenesis of natural silk-polymers in which proteins align (refold), nucleate (denature), and crystallize (aggregate). However for silks, fibrils persist without the need for a drop in temperature.

From a macromolecular perspective the two processes are thought to be similar due to a native protein's unique interaction with its closely bound water. Much like an individual polymer chain in a melt, a native protein and its closely bound water molecules may be considered not as a solution but as a single processable entity, a nanocomposite termed an "aquamelt".

The differences between a typical polymer and an aquamelt are highlighted by an aquamelt's ability to solidify in response to stress at environmental temperatures. This occurs when the stress applied is sufficient to separate the closely bound water from the protein, splitting the nanocomposite. This results in conformational changes to the protein and an increased probability to form hydrogen bonding between protein chains and subsequent solidification. Multiscale structures, i.e., fibrils or foams are the result of a combination of directional stress fields and the self-assembly properties of the aquamelt.

Potential Uses

Aquamelts offer several advantages over current solutions to synthetic polymer production. Firstly they are naturally sourced, with no reliance on oil for production and are recyclable and biodegradable. Secondly they can be processed at room temperatures and pressures resulting in only water as a by-product from the solidification process. Thirdly work calculations performed on silk and high-density polyethylene feedstocks revealed a tenfold difference in the amount of shear energy required in order to initiate solidification. When processing temperature is taken into account the difference in energy requirements to undergo solidification is a thousandfold less for aquamelts than synthetic polymers.

Nanotechnology for Communications

Nanotechnology has been at play in the development of electronics for nearly a decade. Minimum feature sizes in computer chips today are on the order of 50 nm. However exciting advances in this arena are leading to revolutionary strategies for electronic devices. Flexible circuits are folded up into the interior of digital cameras. A new field has arisen in which light is used as the basis of 'plasmonic' devices. Tiny resonating nano machines may be incorporated into future wireless phones. And the entire concept of how a computer 'computes' is being questioned with the potential of nanotubes, nanowires, and nanofibres and possible electronic components.

Nanoscale structuring of optical devices, such as vertical-cavity surface-emitting lasers (VCSELs), quantum dot lasers, and photonic crystal materials, is leading to additional advances in communications technology.

VCSELs have nanoscale layers of compound semiconductors epitaxially grown into their structure—alternating dielectric layers as mirrors and quantum wells. Quantum wells allow the charge carriers to be confined in well-defined regions and provide the energy conversion into light at desired wavelengths. They are placed in the laser's cavity to confine carriers at the nodes of a standing wave and to tailor the band structure for more efficient radiative recombination. One-dimensional nanotechnology techniques involving precise growth of very thin epitaxial semiconductor layers were developed during the 1990s. Such nanostructuring has enhanced the efficiency of VCSELs and reduced the current required for lasing to start (called the threshold current). Because of improving performance and their compatibility with planar manufacturing technology, VCSELs are fast becoming a preferred laser source in a variety of communications applications.

More recently, the introduction of quantum dots (regions so small that they can be given a single electric charge) into semiconductor lasers has been investigated and found to give additional benefits—both further reductions in threshold current and narrower line widths. Quantum dots further confine the optical emission modes within a very narrow spectrum and give the lowest threshold current densities for lasing achieved to date in VCSELs. The quantum dots are introduced into the laser during the growth of strained layers, by a process called Stransky-Krastanov growth. They arise because of the lattice mismatch stress and surface tension of the growing film. Improvements in ways to control precisely the resulting quantum dots to a more uniform single size are still being sought.

Photonic crystals provide a new means to control the steering and manipulation of photons based on periodic dielectric lattices with repeat dimensions on the order of the wavelength of light. These materials can have very exotic properties, such as not allowing light within certain wavelengths to be propagated in a material based on the particular periodic structure. Photonic lattices can act as perfect wavelength-selective mirrors to reflect back incident light from all orientations. They provide the basis for optical switching, steering, and wavelength separation on unprecedented small scales. The periodic structures required for these artificial crystals can be configured as both two- and three-dimensional lattices. Optical sources, switches, and routers are being considered, with two-dimensional planar geometries receiving the most attention, because of their greater ease of fabrication.

Another potentially important communications application for nanotechnology is microelectromechanical systems (MEMS), devices sized at the micrometre level (millionths of a metre). MEMS are currently poised to have a major impact on communications via optical switching. In the future, electromechanical devices may shrink to nanodimensions to take advantage of the higher frequencies of mechanical vibration at smaller masses. The natural (resonant) frequency of vibration for small mechanical beams increases as their size decreases, so that little power is needed to drive them as oscillators. Their efficiency is rated by a quality factor, known as Q, which is a ratio of the energy stored per cycle versus the energy dissipated per cycle. The higher the Q, the more precise the absolute frequency of an oscillator. The Q is very high for micro- and nanoscale mechanical oscillators, and these devices can reach very high frequencies (up to microwave frequencies), making them potential low-power replacements for electronic-based oscillators and filters.

Mechanical oscillators have been made from silicon at dimensions of 10×100 nanometres, where more than 10 percent of the atoms are less than one atomic distance from the surface. While highly homogeneous materials can be made at these dimensions—for example, single-crystal silicon bars—surfaces play an increasing role at nanoscales, and energy losses increase, presumably because of surface defects and molecular species absorbed on surfaces.

It is possible to envision even higher frequencies, in what might be viewed as the ultimate in nanomechanical systems, by moving from nanomachined structures to molecular systems. As an example, multiwalled carbon nanotubes are being explored for their mechanical properties. When the ends of the outer nanotube are removed, the inner tube may be pulled partway out from the outer tube where van der Waals forces between the two tubes will supply a restoring force. The inner tube can thus oscillate, sliding back and forth inside the outer tube. The resonant frequency of oscillation for such structures is predicted to be above one gigahertz (one billion cycles per second). It is unknown whether connecting such systems to the macro world and protecting them from surface effects will ever be practical.

Nanofabrication

Two very different paths are pursued. One is a top-down strategy of miniaturizing current technologies, while the other is a bottom-up strategy of building ever-more-complex molecular devices atom by atom. Top-down approaches are good for producing structures with long-range order and for making macroscopic connections, while bottom-up approaches are best suited for assembly and establishing short-range order at nanoscale dimensions. The integration of top-down and bottom-up techniques is expected to eventually provide the best combination of tools for nanofabrication. Nanotechnology requires new tools for fabrication and measurement.

Nanotechnology in Wireless Devices

Devices that use wireless communication range from RFID tags to TV receivers, and satellites to mobile phones. The availability of internet access from tablets and mobile phones is growing at an exponential rate, causing increasing demands on the performance of wireless networks and mobile devices.

Some passive components necessary in wireless devices, such as inductors and capacitors, cannot be minimized as quickly as transistors and integrated circuits. This imbalance makes it likely that the performance requirements of mobile handsets will exceed the capabilities of current RF technologies within the next 10 to 15 years.

As well as the global growth in mobile internet access via tablets and smartphones, wireless sensors, health monitoring systems, and other devices dependent on wireless communication are becoming more and more prevalent.

Developing more compact, more efficient, and less expensive wireless communication devices will have significant impact on these areas and enable new solutions for healthcare services, logistics and environmental monitoring.

The Wireless Revolution

According to a report by Cisco, global internet traffic from wireless devices (both Wi-Fi and mobile) will exceed that from wired devices by 2015. The convenience, sophistication and flexibility of wireless technology for consumers is very high, which is driving the increasing adoption of wireless internet access over wired access.

The type of activities that consumers engage in over wireless connections is also changing. Even laptop computers and desktop workstations are now often connected via Wi-Fi, as it is perceived as easier to set up for the end-user.

As mobile devices become more and more capable, the limitations on online activities are receding, and the differences in types of traffic between mobile and desktop use are diminishing.

So far, wireless technology is keeping up - LTE networks and other 4G technologies are fast enough to support streaming of video - but if this is to continue, considerable advances in wireless communication technology will be needed.

Nanotechnology-Enabled Wireless Devices

Tuneable Radio Components

Recently, it has been suggested that graphene nanoelectromechanical (NEMS) resonators could be used as tuneable resonators for radio circuits. It is possible for a graphene strip to withstand extremely high strains, making it possible to tune the resonator's operating frequency over a broad bandwidth by electrical straining.

High Frequency Electronics

According to experts, additional frequency bands will also be required to support the increasing quantity of mobile internet traffic. Presently, there are standardization efforts to introduce new frequencies for mobile data traffic in the high frequency range around 2-5 GHz and up to 60GHz. It is anticipated that devices operating at 60 – 100 GHz will be required in the not too distant future.

These very high frequencies cannot be achieved by silicon-based transistor technologies. Novel solutions are required for electronics that can operate at high frequencies. RF transistors made of graphene are promising alternatives.

Wireless Sensors

the modern world is gradually becoming an intelligent, interactive environment, with novel autonomous sensors with wireless communication links incorporated into everyday objects. For this to become reality, these devices must be compact, cheap, and operated on a small battery or on electricity harvested from light, thermal energy or radio waves.

Innovative nano-enabled sensors integrated with small RF transceivers can be utilised for monitoring soil, air quality or water pollution. RFID tags are simple, passive devices that can be read with a radio frequency transceiver. Fabrication by printing can bring down the cost of wireless sensors and the use of nanoparticle inks will largely improve the performance of the printed devices.

Recent Developments

In 2011, IBM research scientists announced that they successfully created a building block for future wireless devices. In a paper published in Science, IBM researchers announced the fabrication of the first IC from wafer-size graphene, and showed a broadband frequency mixer operating at frequencies up to 10 GHz (10 billion cycles/s).

This graphene-based IC was especially developed for wireless communications and can be used for a wide range of applications. At present day conventional frequencies, transceiver and cell phone signals can be improved enabling phones to work where they cannot. At higher frequencies, medical and military personnel can view hidden weapons and perform medical imaging without the radiation exposure dangers of X-rays.

This novel IC includes a graphene transistor and an inductor pair, integrated compactly on a silicon carbide (SiC) wafer. Advances were made in wafer-scale fabrication procedures that maintain graphene quality, and also enable its integration to other components in a complex circuitry. Here the synthesis of graphene is done by thermal annealing of SiC wafers to form uniform graphene layers on the silicon carbide surface. The fabrication of graphene circuits involves four metal and two oxide layers to form on-chip inductor, graphene transistor and interconnects.

The circuit operates similar to a broadband frequency mixer that produces output signals with mixed frequencies (sum and difference) of the input signals. Mixers are the key components of most electronic communication systems. Frequency mixing up to 10 GHz and excellent thermal stability up to 125°C have been shown with the graphene integrated circuit. The developed fabrication scheme can be applied to other graphene materials including chemical vapor deposited (CVD) graphene films synthesized on metal films, and also comply with optical lithography for less thought put and cost.

Nanotechnology in Information and Communication Technology

Information and communication technology is an important and rapidly growing industrial sector with a high rate of innovation. Enormous progress has been made by making a transition from traditional to nanotechnology electronics. Nanotechnology has created a tremendous change in information and communication technology.

Breakthrough Areas

Breakthrough in information and communication technology due to nanotechnology can happen in two steps. First step is top-down miniaturization approach which will take conventional microstructures across the boundary to nanotechnology. Secondly, in the longer term, bottom-up nanoelectronics and nanosystem engineering will emerge using technologies such as self-organisation process to assemble circuits and systems.

Developments

Development are taking place on ultra-integrated (opto)electronics combined with powerful wireless technology as low-price mass products, ultra miniaturization, the design of innovative sensors, production of cheap and powerful polytronic circuits, novel system architectures using nanotechnology for future DNA computing which is interface to biochemical processes and quantum computing which can solve problems for which there are no efficient classical algorithms. Due to the development of nanoelectronic components, quantum cryptography for military and intelligence applications is emerging.

Memory Storage

Memory storage before the advent of nanotechnology relied on transistors, but now reconfigurable arrays are formed for storing large amount of data in small space. For example, we can expect to see the introduction of magnetic RAMs and resonant tunnel elements in logical circuits in the near future. Every single nanobit of a memory storage device is used for storing information. Molecular electronics based on carbon nanotubes or organic macromolecules will be used.

Semiconductors

Nano amplification and chip embedding is used for building semiconductor devices which can even maintain and neutralize the electric flow. Integrated nanocircuits are used in the silicon chips to reduce the size of the processors. Approaches promising success in the medium term include e.g. rapid single-flux quantum (RSFQ) logic or single electron transistors.

Display and Audio Devices

Picture quality and resolution of display devices has improved with the help of nanotechnology. Nanopixelation of these devices make the picture feel real. Similarly frequency modulation in audio devices has been digitized to billionth bit of signals.

Data Processing and Transmission

In the field of data processing and transmission development of electronic, optical and optoelectronic components are expected to lead to lower cost or more precise processes in the field of manufacturing technology. Development of nanoscale logical and storage components are made for the currently dominant CMOS technology using quantum dots and carbon nanotubes. Photonic crystals have potential for use in purely optical circuits as a basis for future information processing based solely on light (photonics). In molecular electronics, nanotechnology can be used to assemble electronic components with new characteristics at atomic level, with advantages including potentially high packing density. Smaller, faster and better components based on quantum mechanical effects, new architectures and new biochemical computing concept called DNA computing are possible with nanotechnology. The new phenomenon, called the "quantum mirage" effect, may enable data transfer within future nanoscale electronic circuits too small to use wires.

Future Nanotechnology Areas

Nanotechnology is the next industrial revolution and the telecommunications industry will be radically transformed by it in the future. Nanotechnology has revolutionized the telecommunications, computing, and networking industries. The emerging innovation technologies are:

* Nanomaterials with novel optical, electrical, and magnetic properties
* Faster and smaller non-silicon-based chipsets, memory, and processors
* New-science computers based on Quantum Computing
* Advanced microscopy and manufacturing systems
* Faster and smaller telecom switches, including optical switches
* Higher-speed transmission phenomena based on plasmonics and other quantum-level phenomena
* Nanoscale MEMS: micro-electro-mechanical systems

2

Photonic Crystals

Photonic crystals are periodic optical nanostructures that affect the motion of photons in much the same way that ionic lattices affect electrons in solids. Photonic crystals occur in nature in the form of structural colouration and promise to be useful in different forms in a range of applications.

In 1887 the English physicist Lord Rayleigh experimented with periodic multi-layer dielectric stacks, showing they had a photonic band-gap in one dimension. Research interest grew with work in 1987 by Yablonovitch and John on periodic optical structures with more than one dimension, now called photonic crystals.

Photonic crystals can be fabricated for one, two, or three dimensions. One-dimensional photonic crystals can be made of layers deposited or stuck together; two-dimensional ones can be made by drilling holes in a suitable substrate, and three-dimensional ones by, for example, stacking spheres in a matrix and dissolving the spheres. Photonic crystals can in principle find uses wherever light must be manipulated. Existing applications include thin-film optics with coatings for lenses and mirrors and colour changing paints. Two-dimensional photonic-crystal fibres are used in nonlinear devices and to guide exotic wavelengths. Three-dimensional crystals may one day be used in optical computers.

Photonic crystals are composed of periodic dielectric, metallo-dielectric or even superconductor microstructures or nanostructures that affect the propagation of electromagnetic waves (EM) in the same way as the periodic potential in a semiconductor crystal affects the electron motion by defining allowed and forbidden electronic energy bands. Photonic crystals contain regularly repeating regions of high and low dielectric constant. Photons (behaving as waves) propagate through this structure – or not – depending on their wavelength. Wavelengths that are allowed to travel are known as modes; groups of allowed modes form bands. Disallowed bands of wavelengths are called photonic band gaps. This gives rise to distinct optical phenomena such as inhibition of spontaneous emission, high-reflecting omni-directional mirrors and low-loss-waveguiding.

The periodicity of the photonic crystal structure must be around half the wavelength of the electromagnetic waves that are to be diffracted. This is ~200 nm (blue) to 350 nm (red) for photonic crystals operating in the visible part of the spectrum. The repeating regions of high and low dielectric constant must therefore be fabricated at this scale, which is difficult.

History of Photonic Crystals

Although photonic crystals have been studied in one form or another since 1887, the term "photonic crystal" was first used over 100 years later, after Eli Yablonovitch and Sajeev John published two milestone papers on photonic crystals in 1987.

Before 1987, one-dimensional photonic crystals in the form of periodic multi-layer dielectric stacks (such as the Bragg mirror) were studied extensively. Lord Rayleigh started their study in 1887, by showing that such systems have a one-dimensional photonic band-gap, a spectral range of large reflectivity, known as a stop-band. Today, such structures are used in a diverse range of applications; from reflective coatings to enhancing the efficiency of LEDs to highly reflective mirrors in certain laser cavities.

A detailed theoretical study of one-dimensional optical structures was performed by Vladimir P. Bykov, who was the first to investigate the effect of a photonic band-gap on the spontaneous emission from atoms and molecules embedded within the photonic structure. Bykov also speculated as to what could happen if two- or three-dimensional periodic optical structures were used. The concept of three-dimensional photonic crystals was then discussed by Ohtaka in 1979, who also developed a formalism for the calculation of the photonic band structure. However, these ideas did not take off until after the publication of two milestone papers in 1987 by Yablonovitch and John. Both these papers concerned high dimensional periodic optical structures – photonic crystals. Yablonovitch's main motivation was to engineer the photonic density of states, to control the spontaneous emission of materials embedded within the photonic crystal; John's idea was to use photonic crystals to affect the localisation and control of light.

After 1987, the number of research papers concerning photonic crystals began to grow exponentially. However, due to the difficulty of actually fabricating these structures at optical scales, early studies were either theoretical or in the microwave regime, where photonic crystals can be built on the far more readily accessible centimetre scale. (This fact is due to a property of the electromagnetic fields known as scale invariance – in essence, the electromagnetic fields, as the solutions to Maxwell's equations, has no natural length scale, and so solutions for centimetre scale structure at microwave frequencies are the same as for nanometre scale structures at optical frequencies.) By 1991, Yablonovitch had demonstrated the first three-dimensional photonic band-gap in the microwave regime. The structure that Yablonvitch was able to produce involved drilling an array of holes in a transparent material, where the holes of each layer form an inverse diamond structure – today it is known as Yablonovite.

In 1996, Thomas Krauss made the first demonstration of a two-dimensional photonic crystal at optical wavelengths. This opened up the way for photonic crystals to be fabricated in semiconductor materials by borrowing the methods used in the semiconductor industry. Today, such techniques use photonic crystal slabs, which are two dimensional photonic crystals "etched" into slabs of semiconductor; total internal reflection confines light to the slab, and allows photonic crystal effects, such as engineering the photonic dispersion to be used in the slab. Research is underway around the world to use photonic crystal slabs in integrated computer chips, to improve the optical processing of communications both on-chip and between chips.

Although such techniques have yet to mature into commercial applications, two-dimensional photonic crystals have found commercial use in the form of photonic crystal fibres (otherwise known as holey fibres, because of the air holes that run through them). Photonic crystal fibres were first developed by Philip Russell in 1998, and can be designed to possess enhanced properties over (normal) optical fibres.

The study of three-dimensional photonic crystals has proceeded more slowly than their two-dimensional counterparts. This is because of the increased difficulty in fabrication; there was no inheritance of readily applicable techniques from the semiconductor industry for fabricators of three-dimensional photonic crystals to draw on. Attempts have been made, however, to adapt some of the same techniques, and quite advanced examples have been demonstrated, for example in the construction of "woodpile" structures constructed on a planar layer-by-layer basis. Another strand of research has been to try and construct three-dimensional photonic structures from self-assembly – essentially allowing a mixture of dielectric nano-spheres to settle from solution into three-dimensionally periodic structures possessing photonic band-gaps. The first demonstration of such "inverse opal" structures came in 2000 by researchers at the University of Toronto, Canada. The ever expanding field of biomimetics – the study of structures from nature in order to better understand and use them in design – is also helping researchers in photonic crystals. For example, in 2006 a naturally-occurring photonic crystal was discovered in scales of a Brazilian beetle.

Construction Strategies

The fabrication method depends on the number of dimensions that the photonic bandgap must exist in.

One-Dimensional Photonic Crystals

In a one-dimensional photonic crystal, layers of different dielectric constant may be deposited or adhered together to form a band gap in a single direction. A Bragg grating is an example of this type of photonic crystal. One-dimensional photonic crystals can be either isotropic or anisotropic, with the latter having potential use as an optical switch.

Recently, a graphene-based Bragg grating (one-dimensional photonic crystal) has been fabricated and demonstrated its capability for excitation of surface electromagnetic waves in the periodic structure by using 633 nm He-Ne laser as the light source. Besides, a novel type of one-dimensional graphene-dielectric photonic crystal has also been proposed. This structure can act as a far-IR filter and also is capable of supporting low-loss surface plasmons for waveguide and sensing applications.

Two-Dimensional Photonic Crystals

In two dimensions, holes may be drilled in a substrate that is transparent to the wavelength of radiation that the bandgap is designed to block. Triangular and square lattices of holes have been successfully employed. The Holey fibre or photonic crystal fibre can be made by taking cylindrical rods of glass in hexagonal lattice, and then heating and stretching them, the triangle-like airgaps between the glass rods become the holes that confine the modes.

Three-Dimensional Photonic Crystals

There are several structure types that have been constructed:

- *Spheres in a diamond lattice*
- Yablonovite

- *The Woodpile Structure* – "rods" are repeatedly etched using beam lithography, filled in and new material is then deposited thereon, and the process is repeated on the next layer with etched channels that are perpendicular to the layer below, and parallel to and out of phase with the channels two layers below. The process is repeated until the structure is of the desired height. The fill-in material is then dissolved using an agent that can dissolve the fill in material but not the deposition material. It is generally hard to introduce defects into this structure.
- *Inverse opals* or *Inverse Colloidal Crystals*-Spheres (such as polystyrene) can be allowed to deposit into a cubic close packed lattice suspended in a solvent. Then a hardener is introduced which makes a transparent solid out of the volume occupied by the solvent. The spheres are then dissolved using an acid such as Hydrochloric acid.
- *A stack of two-dimensional crystals* – This is a more general class of photonic crystals than Yablonovite, but the original implementation of Yablonovite was created using this method.
- The photonic crystal beamsplitter that we made is a fundamental optical component used to control polarized light," explains Dr Mark Turner from Swinburne University. "Specifically what makes our device unique is its ability to directly work with circular polarization at a microscopic scale."
- Circular polarization—Uses 3D laser nanotechnology to exploit circular polarization to build a microscopic prism that contains in excess of 750,000 polymer nanorods. Light focused on this beamsplitter penetrates or is reflected, depending on polarization.

Fabrication Challenges

The major challenge for higher dimensional photonic crystals is in fabrication of these structures, with sufficient precision to prevent scattering losses blurring the crystal properties and with processes that can be robustly mass-produced. One promising method of fabrication for two-dimensionally periodic photonic crystals is a photonic-crystal fibre, such as a "holey fibre". Using fibre draw techniques developed for communications fibre it meets these two requirements, and photonic crystal fibres are commercially available. Another promising method for developing two-dimensional photonic crystals is the so-called photonic crystal slab. These structures consist of a slab of material (such as silicon) which can be patterned using techniques borrowed from the semiconductor industry. Such chips offer the potential to combine photonic processing with electronic processing on a single chip. For three dimensional photonic crystals, various techniques have been used including photolithography and etching techniques similar to those used for integrated circuits. Some of these techniques are already commercially available. To circumvent nanotechnological methods with their complex machinery, alternate approaches have been followed to grow photonic crystals as self-assembled structures from colloidal crystals.

Mass-scale 3D photonic crystal films and fibres can now be produced using a shear-assembly technique which stacks 200–300 nm colloidal polymer spheres into perfect films of fcc lattice. Because the particles have a softer transparent rubber coating the films can be stretched and molded, tuning the photonic bandgaps and producing striking structural colour effects.

Computing Photonic Band Structure

The photonic band gap (PBG) is essentially the gap between the air-line and the dielectric-line in the dispersion relation of the PBG system. To design photonic crystal systems, it is essential to engineer the location and size of the bandgap; this is done by computational modelling using any of the following methods.

1. Plane wave expansion method
2. Finite element method.
3. Finite difference time domain method
4. Order-n spectral method
5. KKR method
6. Bloch wave – MoM method
7. Construction of the Band Diagram

Essentially these methods solve for the frequencies (normal models) of the photonic crystal for each value of the propagation direction given by the wave vector, or vice-versa. The various lines in the band structure, correspond to the different cases of n, the band index.

The plane wave expansion method can be used to calculate the band structure using an eigen formulation of the Maxwell's equations, and thus solving for the eigen frequencies for each of the propagation directions, of the wave vectors. It directly solves for the dispersion diagram. Electric field strength values can also be calculated over the spatial domain of the problem using the eigen vectors of the same problem. For the picture shown to the right, corresponds to the band-structure of a 1D distributed Bragg reflector (DBR) with air-core interleaved with a dielectric material of relative permittivity 12.25, and a lattice period to air-core thickness ratio (d/a) of 0.8, is solved using 101 planewaves over the first irreducible Brillouin zone.

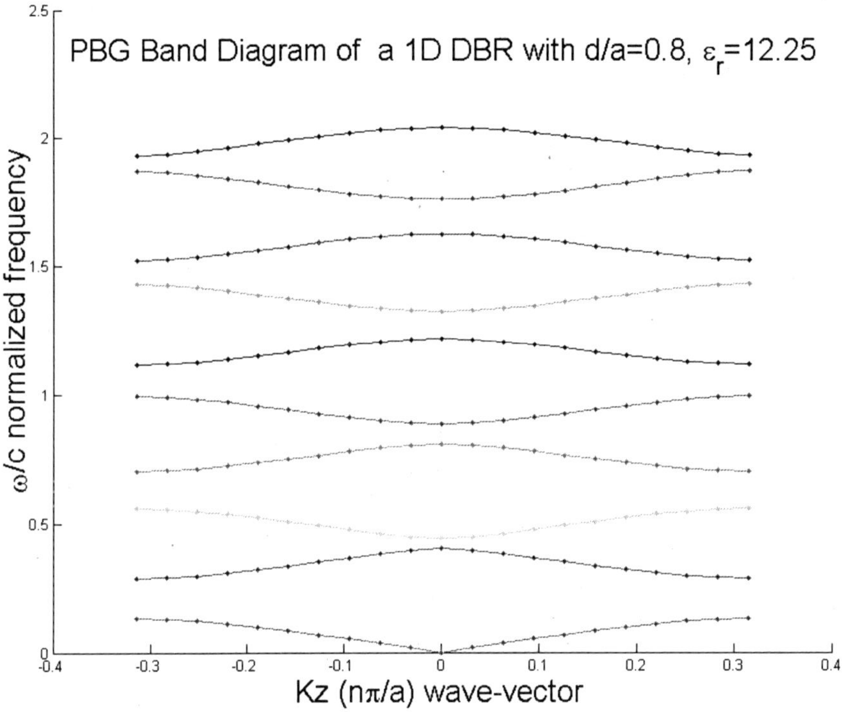

Figure: *Band structure of a 1D Photonic Crystal, DBR air-core calculated using plane wave expansion technique with 101 planewaves, for d/a=0.8, and dielectric contrast of 12.250.*

In order to speed up the calculation of the frequency band structure, the Reduced Bloch Mode Expansion (RBME) method can be used. The RBME method applies "on top" of any of the primary expansion methods mentioned above. For large unit cell models, the RBME method can reduce the time for computing the band structure by up to two orders of magnitude.

Applications

Photonic crystals are attractive optical materials for controlling and manipulating the flow of light. One dimensional photonic crystals are already in widespread use in the form of thin-film optics with applications ranging from low and high reflection coatings on lenses and mirrors to colour changing paints and inks. Higher dimensional photonic crystals are of great interest for both fundamental and applied research, and the two dimensional ones are beginning to find commercial applications. The first commercial products involving two-dimensionally periodic photonic crystals are already available in the form of photonic-crystal fibres, which use a microscale structure to confine light with radically different characteristics compared to conventional optical fibre for applications in nonlinear devices and guiding exotic wavelengths. The three-dimensional counterparts are still far from commercialization but may offer additional features such as optical nonlinearity required for the operation of optical transistors used in optical computers, when some technological aspects such as manufacturability and principal difficulties such as disorder are under control.

Nanocomputer

Nanocomputer is the logical name for a computer smaller than the microcomputer, which is smaller than the minicomputer. (The minicomputer is called "mini" because it was a lot smaller than the original (mainframe) computers.)

The term nanocomputer is increasingly used to refer to general computing devices of size comparable to a credit card. The first use of this name appears to be to describe the s1mp3 manufactured by The Flying Electron Inc. (7 November 2008). It is now used for a wider range of devices, such as

- Arduino
- BeagleBoard
- Olinuxino
- Odroid
- Miracast dongles
- All the PC-on-a-stick (MK802 series...)
- Raspberry Pi

A perhaps more principled interpretation (analogous to the first microcomputers having fundamental parts no bigger than several micrometres) is that of a computer whose fundamental parts are no bigger than a few nanometres. There are several ways nanocomputers might be built, using mechanical, electronic, biochemical, or quantum technology. It was argued in 1998 that is unlikely that nanocomputers will be made out of semiconductor transistors (Microelectronic components that are at the core of all modern electronic devices), as they seem to perform significantly less well when shrunk to sizes under 100 nanometres.

This is rather contradicted by the smallest part of current state-of-the-art microprocessors measuring 22 nm as of April 2012. Moreover Intel's 5 nanometre technology outlook predicts 5 nm feature size by

2022. The International Technology Roadmap for Semiconductors gives an industrial consensus on feature scaling following Moore's Law.

Note that a Silicon-Silicon bond length is 235.2 pm, which means that a 5 nm-width transistor would be 21 silicon atoms wide.

Nanochip

"Nanochip" is a term referring either to a typical computer chip, which has feature sizes measured in nanometres, or some novel computer design based on "nanotechnology," another vaguely defined word. Basically, the word is vague enough that you can theoretically use it to refer to any modern computer chip.

More specifically, the term "Nanochip" is most often associated with Nanochip, Inc., a company attempting to use MEMS (micro-electro-mechanical systems) technology to create a high-density storage device that leapfrogs improvements in traditional storage methods, which are based on photolithography. Nanochip's technology uses arrays of tiny atomic force probe tips to electrically write to a storage medium. The probe tips are only 25 nm across, and can edit areas of the storage media with a precision of 10 nanometres or less. According to the company, this could enable microchips with 100 GB storage capacity by 2009, and terabytes (TB) further down the line.

Nanochip, Inc. is backed by various VC firms and larger chip companies including Intel. Its array probe technology is reminiscent of IBM's Millipede, a defunct project that used a similar approach. Millipede was apparently abandoned sometime around 2006, for unknown reasons. Meanwhile, Nanochip has been working on its technology since 1996, but only began receiving media coverage around 2007. It remains to be seen whether Nanochip's technology will see the light of day or whether it will fail like Millipede.

Sometimes, the word "nanochip" is also used to refer to conventional computer chips that integrate carbon nanowires into their structure. Carbon nanowires are highly conductive and chips that use them cannot be manufactured exclusively using traditional photolithography — they must go through two manufacturing stages, one for the conventional chip and one to add on the nanowires.

Nanowires are useful for computing because of their very small size and high conductivity. Sometime between 2015 and 2025, it seems feasible that computers will be built almost exclusively out of carbon nanowires, but for now, we're in a transitional period between chips manufactured using photolithography and chips made with more advanced methods.

Energy Applications of Nanotechnology

Over the past few decades, the fields of science and engineering have been seeking to develop new and improved types of energy technologies that have the capability of improving life all over the world. In order to make the next leap forward from the current generation of technology, scientists and engineers have been developing energy applications of nanotechnology. Nanotechnology, a new field in science, is any technology that contains components smaller than 100 nanometres. For scale, a single virus particle is about 100 nanometres in width.

An important subfield of nanotechnology related to energy is nanofabrication. Nanofabrication is the process of designing and creating devices on the nanoscale. Creating devices smaller than 100 nanometres opens many doors for the development of new ways to capture, store, and transfer energy. The inherent level of control that nanofabrication could give scientists and engineers would be critical in providing the capability of solving many of the problems that the world is facing today related to the current generation of energy technologies.

People in the fields of science and engineering have already begun developing ways of utilising nanotechnology for the development of consumer products. Benefits already observed from the design of these products are an increased efficiency of lighting and heating, increased electrical storage capacity, and a decrease in the amount of pollution from the use of energy. Benefits such as these make the investment of capital in the research and development of nanotechnology a top priority.

Consumer Products

Recently, previously established and entirely new companies such as BetaBatt, Inc. and Oxane Materials are focusing on nanomaterials as a way to develop and improve upon older methods for the capture, transfer, and storage of energy for the development of consumer products.

ConsERV, a product developed by the Dais Analytic Corporation, uses nanoscale polymer membranes to increase the efficiency of heating and cooling systems and has already proven to be a lucrative design. The polymer membrane was specifically configured for this application by selectively engineering the size of the pores in the membrane to prevent air from passing, while allowing moisture to pass through the membrane. ConsERV's value is demonstrated in the form of an energy recovery a device which pretreats the incoming fresh air to a building using the energy found in the exhaust air steam using no moving parts to lower the energy and carbon footprint of existing forms of heating and cooling equipment Polymer membranes can be designed to selectively allow particles of one size and shape to pass through while preventing others of different dimensions. This makes for a powerful tool that can be used in all markets - consumer, commercial, industrial, and government products from biological weapons protection to industrial chemical separations. Dais's near term uses of this 'family' of selectively engineered nanotechnology materials, aside from ConsERV, include (a.) a completely new cooling cycle capable of replacing the refrigerant based cooling cycle the world has known for the past 100 plus years. This product, under development, is named NanoAir. NanoAir uses only water and this selectively engineered membrane material to cool (or heat) and dehumidify (or humidify) air. There are no fluorocarbon producing gasses used, and the energy required to cool a space drops as thermodynamics does the actual cooling. The company was awarded an Advanced Research Program Administration - Energy award in 2010, and a United States Department of Defence (DoD) grant in 2011 both designed to accelerate this newer, energy efficient technology closer to commercialization, and (b.) a novel way to clean most all contaminated forms of water called NanoClear. By using the selectivity of this hermetic, engineered composite material it can transfer only a water molecule from one face of the membrane to the other leaving behind the contaminants. It should also be noted Dais received a US Patent (Patent Number 7,990,679) in October 2011 titled "Nanoparticle Ultracapacitor". This patented item again uses the selectively engineered material to create an energy storage mechanism projected to have performance and cost advantages over existing storage technologies. The company has used this patent's concepts to create a functional energy storage prototype device named NanoCap. NanoCap is a form of ultra-capacitor potentially useful to power a broad range of applications including most forms of transportation, energy storage (especially useful as a storage media for renewable energy technologies), telecommunication infrastructure, transistor gate dielectrics, and consumer battery applications (cell phones, computers, etc.).

A New York based company called Applied NanoWorks, Inc. has been developing a consumer product that utilises LED technology to generate light. Light-emitting diodes or LEDs, use only about 10% of the energy that a typical incandescent or fluorescent light bulb uses and typically last much longer, which makes them a viable alternative to traditional light bulbs. While LEDs have been around for decades, this company and others like it have been developing a special variant of LED called the white LED. White LEDs consist of semi-conducting organic layers that are only about 100 nanometres in distance from each other and are placed between two electrodes, which create an anode, and a cathode. When voltage is applied to the system, light is generated when electricity passes through the two organic layers. This is called electroluminescence. The semiconductor properties of the organic layers are what allow for the minimal amount of energy necessary to generate light. In traditional light bulbs, a metal filament is used to generate light when electricity is run through the filament. Using metal generates a great deal of heat and therefore lowers efficiency.

Research for longer lasting batteries has been an ongoing process for years. Researchers have now begun to utilise nanotechnology for battery technology. mPhase Technologies in conglomeration with Rutgers University and Bell Laboratories have utilised nanomaterials to alter the wetting behaviour of the surface where the liquid in the battery lies to spread the liquid droplets over a greater area on the surface and therefore have greater control over the movement of the droplets. This gives more control to the designer of the battery. This control prevents reactions in the battery by separating the electrolytic liquid from the anode and the cathode when the battery is not in use and joining them when the battery is in need of use.

Thermal applications also are a future applications of nanothechonlogy creating low cost system of heating, ventilation, and air conditioning, changing molecular structure for better management of temperature

Reduction of Energy Consumption

A reduction of energy consumption can be reached by better insulation systems, by the use of more efficient lighting or combustion systems, and by use of lighter and stronger materials in the transportation sector. Currently used light bulbs only convert approximately 5% of the electrical energy into light. Nanotechnological approaches like or quantum caged atoms (QCAs) could lead to a strong reduction of energy consumption for illumination.

Increasing the Efficiency of Energy Production

Today's best solar cells have layers of several different semiconductors stacked together to absorb light at different energies but they still only manage to use 40 percent of the Sun's energy. Commercially available solar cells have much lower efficiencies (15-20%). Nanotechnology could help increase the efficiency of light conversion by using nanostructures with a continuum of bandgaps.

The degree of efficiency of the internal combustion engine is about 30-40% at present. Nanotechnology could improve combustion by designing specific catalysts with maximized surface area. In 2005, scientists at the University of Toronto developed a spray-on nanoparticle substance that, when applied to a surface, instantly transforms it into a solar collector.

Nuclear Accident Cleanup and Waste Storage

Nanomaterials deployed by swarm robotics may be helpful for decontaminating a site of a nuclear accident which poses hazards to humans because of high levels of radiation and radioactive particles. Hot nuclear compounds such as corium or melting fuel rods may be contained in "bubbles" made from nanomaterials that are designed to isolate the harmful effects of nuclear activity occurring inside of them from the outside environment that organisms inhabit.

Economic Benefits

The relatively recent shift toward using nanotechnology with respect to the capture, transfer, and storage of energy has and will continue to have many positive economic impacts on society. The control of materials that nanotechnology offers to scientists and engineers of consumer products is one of the most important aspects of nanotechnology. This allows for an improved efficiency of products across the board.

A major issue with current energy generation is the loss of efficiency from the generation of heat as a by-product of the process. A common example of this is the heat generated by the internal combustion engine. The internal combustion engine loses about 64% of the energy from gasoline as heat and an improvement of this alone could have a significant economic impact. However, improving the internal combustion engine in this respect has proven to be extremely difficult without sacrificing performance. Improving the efficiency of fuel cells through the use of nanotechnology appears to be more plausible by using molecularly tailored catalysts, polymer membranes, and improved fuel storage.

In order for a fuel cell to operate, particularly of the hydrogen variant, a noble-metal catalyst (usually platinum, which is very expensive) is needed to separate the electrons from the protons of the hydrogen atoms. However, catalysts of this type are extremely sensitive to carbon monoxide reactions. In order to combat this, alcohols or hydrocarbons compounds are used to lower the carbon monoxide concentration in the system. This adds an additional cost to the device. Using nanotechnology, catalysts can be designed through nanofabrication that are much more resistant to carbon monoxide reactions, which improves the efficiency of the process and may be designed with cheaper materials to additionally lower costs. Fuel cells that are currently designed for transportation need rapid start-up periods for the practicality of consumer use. This process puts a lot of strain on the traditional polymer electrolyte membranes, which decreases the life of the membrane requiring frequent replacement. Using nanotechnology, engineers have the ability to create a much more durable polymer membrane, which addresses this problem. Nanoscale polymer membranes are also much more efficient in ionic conductivity. This improves the efficiency of the system and decreases the time between replacements, which lowers costs.

Another problem with contemporary fuel cells is the storage of the fuel. In the case of hydrogen fuel cells, storing the hydrogen in gaseous rather than liquid form improves the efficiency by 5%. However, the materials that we currently have available to us significantly limit fuel storage due to low stress tolerance and costs. Scientists have come up with an answer to this by using a nanoporous styrene material (which is a relatively inexpensive material) that when super-cooled to around -196°C, naturally holds on to hydrogen atoms and when heated again releases the hydrogen for use.

Capacitors: Then and Now

For decades, scientists and engineers have been attempting to make computers smaller and more efficient. A crucial component of computers are capacitors. A capacitor is a device that is made of a pair of electrodes separated by an insulator that each stores an opposite charge. A capacitor stores a charge when it is removed from the circuit that it is connected to; the charge is released when it is replaced back into the circuit. Capacitors have an advantage over batteries in that they release their charge much more quickly than a battery.

Traditional or foil capacitors are composed of thin metal conducting plates separated by an electrical insulator, which are then stacked or rolled and placed in a casing. The problem with a traditional capacitor such as this is that they limit how small an engineer can design a computer. Scientists and engineers have since turned to nanotechnology for a solution to the problem.

Using nanotechnology, researchers developed what they call "ultracapacitors." An ultracapacitor is a general term that describes a capacitor that contains nanocomponents. Ultracapacitors are being researched heavily because of their high density interior, compact size, reliability, and high capacitance. This decrease in size makes it increasingly possible to develop much smaller circuits and computers.

Ultracapacitors also have the capability to supplement batteries in hybrid vehicles by providing a large amount of energy during peak acceleration and allowing the battery to supply energy over longer periods of time, such as during a constant driving speed. This could decrease the size and weight of the large batteries needed in hybrid vehicles as well as take additional stress off the battery. However, the combination of ultracapacitors and a battery is not cost effective due to the need of additional DC/DC electronics to coordinate the two.

Nanoporous carbon aerogel is one type of material that is being utilised for the design of ultracapacitors. These aerogels have a very large interior surface area and can have its properties altered by changing the pore diameter and distribution along with adding nanosized alkali metals to alter its conductivity.

Carbon nanotubes are another possible material for use in an ultracapacitor. Carbon nanotubes are created by vaporizing carbon and allowing it to condense on a surface. When the carbon condenses, it forms a nanosized tube composed of carbon atoms. This tube has a high surface area, which increases the amount of charge that can be stored. The low reliability and high cost of using carbon nanotubes for ultracapacitors is currently an issue of research.

In a study concerning ultracapacitors or supercapacitors, researchers at the Sungkyunkwan University in the Republic of Korea explored the possibility of increasing the capacitance of electrodes through the addition of fluorine atoms to the walls of carbon nanotubes. As briefly mentioned before, carbon nanotubes are an increasing form of capacitors due to their superb chemical stability, high conductivity, light mass, and their large surface area. These researchers fluorinated single-walled carbon nanotubes (SWCNTs) at high temperatures to bind fluorine atoms to the walls. The attached fluorine atoms changed the non-polar nanotubes to become polar molecules.

This can be attributed to the charge transfer from the fluorine. This created dipole-dipole layers along the carbon nanotube walls. Testing of these fluorinated SWCNTs against normal state SWCNTs showed a difference in capacitance. It was determined that the fluorinated SWCNTs are advantageous in fabricating electrodes for capacitors and improve the wettability with aqueous electrolytes, which promotes the overall performance of supercapacitors. While this study brought to knowledge a more efficient example of capacitors, little is known about this new supercapacitor, large scale synthesis is lacking and is necessary for any massive production, and preparation conditions are quite tedious in achieving the final product.

Theory of Capacitance

Understanding the concept of capacitance can be helpful in understanding why nanotechnology is such a powerful tool for the design of higher energy storing capacitors. A capacitor's capacitance (C) or amount of energy stored is equal to the amount of charge (Q) stored on each plate divided by the voltage (V) between the plates. Another representation of capacitance is that capacitance (C) is approximately equal to the permittivity (ε) of the dielectric times the area (A) of the plates divided by the distance (d) between them. Therefore, capacitance is proportional to the surface area of the conducting plate and inversely proportional to the distance between the plates.

Using carbon nanotubes as an example, a property of carbon nanotubes is that they have a very high surface area to store a charge. Using the above proportionality that capacitance (C) is proportional to the surface area (A) of the conducting plate; it becomes obvious that using nanoscaled materials with high surface area would be great for increasing capacitance. The other proportionality described above is that capacitance (C) is inversely proportional to the distance (d) between the plates. Using nanoscaled plates such as carbon nanotubes with nanofabrication techniques, gives the capability of decreasing the space between plates which again increases capacitance.

Nanobatteries are fabricated batteries employing technology at the nanoscale, a scale of minuscule particles that measure less than 100 nanometres or $100x10^{-9}$ metres. In comparison, traditional Li-Ion technology uses active materials, such as cobalt-oxide or manganese oxide, with particles that range in size between 5 and 20 micrometres (5000 and 20000 nanometres - over 100 times nanoscale). It is hoped that nano-engineering will improve many of the shortcomings of present battery technology, such as recharging time and battery 'memory'.

Several companies are researching and developing these technologies. In March 2005, Toshiba announced that they had a new Lithium-Ion battery with a nanostructured lattice at the cathode and anode that allowed the battery to recharge a surprising eighty times faster than previously. Prototype models were able to charge to eighty percent capacity in one minute, and were one hundred percent recharged after 10 minutes.

When a traditional lithium-ion battery is charged too quickly, it creates a bottleneck in which the lithium moving through electrolyte liquid from the negative electrode to the positive backs up on the surface of the liquid. Under slower charging conditions, the lithium "hides" in void space and does not cause a problem.

"Liquid electrolyte is unstable in the presence of metallic lithium and will cause all sorts of problems. That is why it is imperative to observe the slow-charging rate rule with lithium-ion batteries," Donald Sadoway, MIT professor of materials chemistry and an electrochemistry researcher, explained to TechNewsWorld. Sadoway said the consequences could be as severe as the battery exploding.

Background

Nanobatteries are generally described by three sections

- Anode
- Cathode
- Electrolyte

In Lithium ion batteries the anode is almost always graphite, so most research is being done on the cathode and electrolyte materials. By reducing the size of the materials used in a nanobattery, higher conductivity can be reached, leading to an increase in power, in both charge and discharge.

Advantages

Using nanotechnology in the manufacture of batteries offers the following benefits:

- Increasing the available power from a battery and decreasing the time required to recharge a battery. These benefits are achieved by coating the surface of an electrode with nanoparticles. This increases the surface area of the electrode thereby allowing more current to flow between the electrode and the chemicals inside the battery. This technique could increase the efficiency of hybrid vehicles by significantly reducing the weight of the batteries needed to provide adequate power.
- Increasing the shelf life of a battery by using nanomaterials to separate liquids in the battery from the solid electrodes when there is no draw on the battery. This separation prevents the low level discharge that occurs in a conventional battery, which increases the shelf life of the battery dramatically.
- Reducing the possibility of batteries catching fire by providing less flammable electrode material.

Unique Fabrication

A gel is created by the chosen design and is used to impregnate the anodic pores. This impregnated gel is placed between membrane walls and on top of the electrolyte. The anodic film is placed below the electrolyte, with the width of the cathode being much smaller than the height of the cathode. The separation of the cathodes by the membrane walls creates in essence a series of nanobatteries. This is helpful in research because it allows each set of cathode to be tested separately, or all at once.

Academic Research

In 2007, the first cross-sectional observation of an all solid state Li-ion nanobattery was taken by TEM. By looking at a nanobattery in TEM, the deterioration on the battery interface due to cycling is able to be observed with an attempt to not only understand by finding the underlying causes behind battery deterioration. The next step in the process is to cycle the battery while in TEM so that the live deterioration can be observed. Three layers of the battery were looked at in TEM, with two nanobatteries observed. The first nanobattery was pristine and uncycled, while the second nanobattery was run through ten cycles so that the deterioration might be characterized. A large irreversible capacity between the first charge and discharge was seen in the cycled nanobattery. The capacity of the battery was also seen to disappear rapidly during cycling.

Printable Batteries

Researchers at the University of California, Los Angeles have successfully developed a "nanotube ink" for manufacturing flexible batteries using printed electronics techniques. Using nanotube ink, the carbon cathode tube and manganese oxide electrolyte components of a zinc-carbon battery can be printed as different layers on a surface, over which an anode layer of zinc foil can be printed. The resultant battery is less than a millimetre thick. Although discharge currents of the batteries are at present below the level of practical use, the nanotubes in the ink allow the charge to conduct more efficiently than in a conventional battery, such that the nanotube technology could lead to improvements in battery performance.

Researching Companies

Toshiba

By using nanomaterial, Toshiba has increased the surface area of the lithium and widened the bottleneck, allowing the particles to pass through the liquid and recharge the battery more quickly. Toshiba states that it tested a new battery by discharging and fully recharging one thousand times at 77 degrees and found that it lost only one percent of its capacity, an indication of a long battery life.

Toshiba's battery is 3.8 mm thick, 62 mm high and 35 mm deep.

A123Systems

A123Systems has also developed a commercial nano Li-Ion battery. A123 Systems claims their battery has the widest temperature range at -30 .. +70 p C. Much like Toshiba's nanobattery, A123 Li-Ion batteries charge to "high capacity" in five minutes. Safety is a key feature touted by the A123 technology, with a video on their website of a nail drive test, in which a nail is driven through a traditional Li-Ion battery and an A123 Li-Ion battery, where the traditional battery flames up and bubbles at one end, the A123 battery simply emits a wisp of smoke at the penetration site. Thermal conductivity is another selling point for the A123 battery, with the claim that the A123 battery offers 4 times higher thermal conductivity than conventional Lithium-Ion cylindrical cells. The nanotechnology they employ is a patented nanophosphate technology.

Valence

Also in the market is Valence Technology, Inc. The technology they are marketing is Saphion Li-Ion Technology. Like A123, they are using a nanophosphate technology, and different active materials than traditional Li-Ion batteries.

Altair

AltairNano has also developed a nanobattery with a one-minute recharge. The advance that Altair claims to have made is in the optimization of nano-structured lithium titanate spinel oxide (LTO).

U.S. Photonics

U.S. Photonics is in the process of developing a nanobattery utilising "environmentally friendly" nanomaterials for both the anode and cathode as well as arrays of individual nano-sized cell containers for the solid polymer electrolite. U.S. Photonics has recently received a National Science Foundation SBIR phase I grant for development of nanobattery technology.

Next Alternative Inc.

Next Alternative has a new Carbon Nanotube (CNT) battery that is a modification of existing car battery types that will allow for the battery to recharge in less than 10 minutes and has a Reserve Capacity of at least 8 times the original unmodified battery. The major difference comes from a typical lead acid battery providing 12-15 kW-hours of electricity or a range of 50–100 miles, where the CNT lead/lead-acid battery will deliver 380 miles distance between charges. This battery could also be recharged in under 10 minutes. The typical lead-acid battery has a recharge time between 4 and 10 hours. The recharge life of the battery (200 cycles for lead-acid) can be extended by at a minimum of 4 times with the new CNT lead/lead-acid battery.

Sony

Produced the first cobalt-based lithium-ion battery in 1991. Since the inception of this first Li-ion battery, the research of nanobatteries has been underway with Sony continuing their strides into the nanobattery field.

Nanotechnology in Solar Cells

How can Nanotechnology Improve Solar Cells

Using nanoparticles in the manufacture of solar cells has the following benefits:

- Reduced manufacturing costs as a result of using a low temperature process similar to printing instead of the high temperature vacuum deposition process typically used to produce conventional cells made with crystalline semiconductor material.
- Reduced installation costs achieved by producing flexible rolls instead of rigid crystalline panels. Cells made from semiconductor thin films will also have this characteristic.
- Currently available nanotechnology solar cells are not as efficient as traditional ones, however their lower cost offsets this. In the long term nanotechnology versions should both be lower cost and, using quantum dots, should be able to reach higher efficiency levels than conventional ones.

Solar Cells: Nanotechnology Applications under Development

Researchers at Michigan Technological University have developed a honeycomb like structure of graphene in which the graphene sheets are held apart by lithium carbonate. They have used this “3D graphene” to replace the platinum in a dye sensitized solar cell and achieved 7.8 percent conversion of sunlight to electricity.

Researchers at Los Alamos National Lab have demonstrated a solar cell that uses a copper indium selenide sulfide quantum dots. Unlike quantum dots containing lead or cadium the copper based quantum dot is non-toxic as well as low cost.

Researchers at MIT are studying solar cells made from single molecule thick sheets of graphene and materials such as molybdenum diselenide. They are predicting that this type of solar cells could produce up to 1000 times as much more power for a given weigh of material than conventional solar cells. They have completed computer modelling and are working on building the solar cells.

Researchers at MIT have developed a solar cell using graphene coated with zinc oxide nanowires. The researchers believe that this method will allow the production of low cost flexible solar cells at high enough efficiency to be competive.

Researchers are using a method called Aerotaxy to grow semiconducting nanowires on gold nanoparticles. They plan to use self assembly techniques to align the nanowires on a substrate; forming a solar cell or other electrical devices.

The gold nanoparticles replace the silicon substrate on which conventional semiconductor based solar cells are built. Researchers at Princeton University have developed a solar cell that almost eliminates losses due to reflection of light. They use a 30 nanometre thick gold nanomesh (a layer with a regular pattern of 175 nanometre diameter holes) along with an active layer thinner than the wavelength of light. They found that this combination traps most of the light in the solar cell, increasing the efficiency of the cell.

Researchers at Duke University are developing another method to reduce losses due to the reflection of light. In this method the combination of silver nanocubes scattered over a thin gold layer reduce losses due to reflection.

Researchers at the Nies Bohr Institute have determined that sunlight can be concentrated in nanowires due to a resonance effect. This effect can result in more efficient solar cells, allowing more of the energy from the sun to be converted to electricity.

An inexpensive solar cell designed to produce hydrogen is being developed using iron oxide nanoparticles. Combining silver nanowires, titanium dioxide nanoparticles and a polymer that absorbs infrared light to make a solar cell that is about 70 percent transparent to visible light, allowing it to be used in windows.

Titanium dioxide nanotubes filled with a polymer to form low cost solar cells

Combining lead selenide quantum dots with titanium dioxide to form higher efficiency solar cells.

Combining carbon nanotubes and buckyballs to produce solar cells. Some researchers combine the nanotubes and buckyballs with a polymer, while another group of researchers are only using nanotubes and buckyballs. A third research group is also using nanotubes and buckballs along with graphene to build a solar cell.

Researchers at Stanford University have found a way to trap light in organic solar cells. The idea is that the longer light is in the solar cell the more electrons will be generated. The researchers found that by making the organic layer much thinner than the wavelength of light and sandwiching the organic layer between a mirror layer and a rough layer the light stayed in the solar cell longer and excited more electrons.

Semiconductor nanoparticles applied in a low temperature printing process that results in low cost solar cells.

Organic molecules to lower costs.

Using light absorbing nanowires embedded in a flexible polymer film is another method being developed to produce low cost flexible solar panels.

Using light absorbing graphene sheets to produce low cost solar panels

Organic solar cells that are self repairing

Organic solar cells that can be applied by spray painting, possibly turning the surface of a car into a solar cell.

Researchers at Lawrence Berkeley have demonstrated an inexpensive process for making solar cells. These solar cells are composed of cadmium sulfide nanowires coated with copper sulfide.

Solar cells that can be installed as a coating on windows or other building materials, referred to as "Building Integrated Photovoltaic's".

3

Colloidal Gold

Colloidal gold is a suspension (or colloid) of submicrometre-size particles of gold in a fluid, usually water. The liquid is usually either an intense red colour (for particles less than 100 nm) or blue/purple (for larger particles). Due to the unique optical, electronic, and molecular-recognition properties of gold nanoparticles, they are the subject of substantial research, with applications in a wide variety of areas, including electron microscopy, electronics, nanotechnology, and materials science.

Properties and applications of colloidal gold nanoparticles strongly depend upon their size and shape. For example, rodlike particles have both transverse and longitudinal absorption peak, and anisotropy of the shape affects their self-assembly.

Known, or at least used (perhaps proceeding by accident without much understanding of the process) since ancient times, the synthesis of colloidal gold was crucial to the 4th-century Lycurgus Cup, which changes colour depending on the direction of the light. Later it was used as a method of staining glass.

A so-called Elixir of Life, a potion made from gold, was discussed, if not actually manufactured, in ancient times. In the 16th century, the alchemist Paracelsus claimed to have created a potion called *Aurum Potabile* (Latin: potable gold). In *The newe iewell of health*, 1576, translator George Baker promoted the use and preparation of potable gold, along with other "most excellent secretes of phisicke and philosophie".

In the 17th century, the glass-colouring process was refined by Andreus Cassius and Johann Kunckel, allowing them to produce a striking deep-ruby coloured form of glass. In 1842, John Herschel invented a photographic process called chrysotype (from the Greek ÷ñáóüò meaning "gold") that used colloidal gold to record images on paper.

Modern scientific evaluation of colloidal gold did not begin until Michael Faraday's work of the 1850s. Paracelsus' work is known to have inspired Faraday to prepare the first pure sample of colloidal gold, which he called 'activated gold', in 1857. He used phosphorus to reduce a solution of gold chloride.

For a long time, the composition of the Cassius ruby-gold was unclear. Several chemists suspected it to be a gold tin compound, due to its preparation. Faraday was the first to recognise that the colour was actually due to the minute size of the gold particles. In 1898, Richard Adolf Zsigmondy prepared the first colloidal gold in diluted solution. Apart from Zsigmondy, Theodor Svedberg, who invented ultracentrifugation, and Gustav Mie, who provided the theory for scattering and absorption by spherical particles, were also interested in understanding synthesis and properties of colloidal gold.

Synthesis

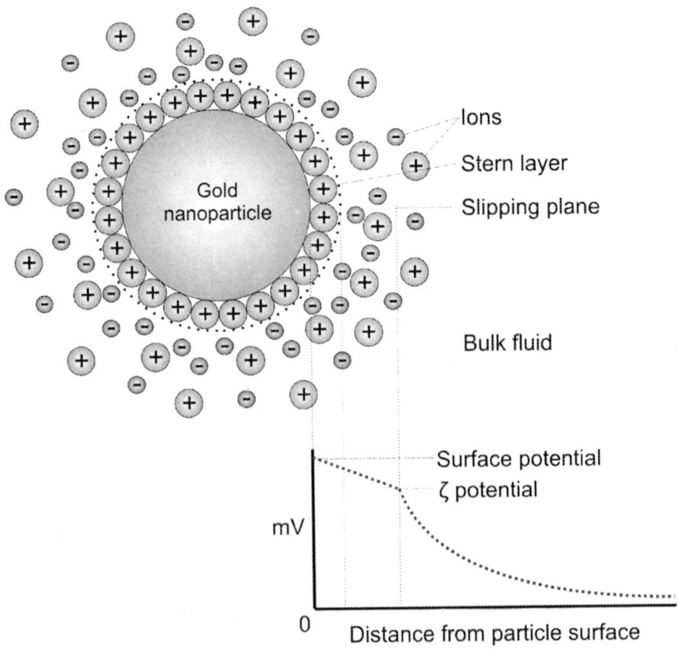

Figure: *Potential difference as a function of distance from particle surface.*

Generally, gold nanoparticles are produced in a liquid ("liquid chemical methods") by reduction of chloroauric acid ($H[AuCl_4]$). After dissolving $H[AuCl_4]$, the solution is rapidly stirred while a reducing agent is added. This causes Au^{3+} ions to be reduced to neutral gold atoms. As more and more of these gold atoms form, the solution becomes supersaturated, and gold gradually starts to precipitate in the form of sub-nanometre particles. The rest of the gold atoms that form stick to the existing particles, and, if the solution is stirred vigorously enough, the particles will be fairly uniform in size.

To prevent the particles from aggregating, some sort of stabilizing agent that sticks to the nanoparticle surface is usually added. Also, gold colloids can be synthesised without stabilizers by laser ablation in liquids. They can be functionalized with various organic ligands to create organic-inorganic hybrids with advanced functionality.

Turkevich Method

The method pioneered by J. Turkevich et al. in 1951 and refined by G. Frens in 1970s, is the simplest one available. In general, it is used to produce modestly monodisperse spherical gold nanoparticles suspended in water of around 10–20 nm in diameter. Larger particles can be produced, but this comes at the cost of monodispersity and shape. It involves the reaction of small amounts of hot chlorauric acid with small amounts of sodium citrate solution. The colloidal gold will form because the citrate ions act as both a reducing agent and a capping agent.

Recently, the evolution of the spherical gold nanoparticles in the Turkevich reaction has been elucidated. It is interesting to note that extensive networks of gold nanowires are formed as a transient intermediate. These gold nanowires are responsible for the dark appearance of the reaction solution before it turns ruby-red.

To produce larger particles, less sodium citrate should be added (possibly down to 0.05%, after which there simply would not be enough to reduce all the gold). The reduction in the amount of sodium citrate will reduce the amount of the citrate ions available for stabilizing the particles, and this will cause the small particles to aggregate into bigger ones (until the total surface area of all particles becomes small enough to be covered by the existing citrate ions).

Brust Method

This method was discovered by Brust and Schiffrin in early 1990s, and can be used to produce gold nanoparticles in organic liquids that are normally not miscible with water (like toluene). It involves the reaction of a chlorauric acid solution with tetraoctylammonium bromide

(TOAB) solution in toluene and sodium borohydride as an anti-coagulant and a reducing agent, respectively.

Here, the gold nanoparticles will be around 5–6 nm. $NaBH_4$ is the reducing agent, and TOAB is both the phase transfer catalyst and the stabilizing agent.

It is important to note that TOAB does not bind to the gold nanoparticles particularly strongly, so the solution will aggregate gradually over the course of approximately two weeks. To prevent this, one can add a stronger binding agent, like a thiol (in particular, alkanethiols), which will bind to gold, producing a near-permanent solution. Alkanethiol protected gold nanoparticles can be precipitated and then redissolved. Some of the phase transfer agent may remain bound to the purified nanoparticles, this may affect physical properties such as solubility. In order to remove as much of this agent as possible the nanoparticles must be further purified by soxhlet extraction.

Perrault Method

This approach, discovered by Perrault and Chan in 2009, uses hydroquinone to reduce $HAuCl_4$ in an aqueous solution that contains gold nanoparticle seeds. This seed-based method of synthesis is similar to that used in photographic film development, in which silver grains within the film grow through addition of reduced silver onto their surface. Likewise, gold nanoparticles can act in conjunction with hydroquinone to catalyze reduction of ionic gold onto their surface. The presence of a stabilizer such as citrate results in controlled particle growth. Typically, the nanoparticle seeds are produced using the citrate method. The hydroquinone method complements that of Frens, as it extends the range of monodispersed spherical particle sizes that can be produced. Whereas the Frens method is ideal for particles of 12-20 nm, the hydroquinone method can produce particles of at least 30–250 nm.

Navarro Method

As pointed out before, the precise size control with a low polydispersity of spherical gold nanoparticles remains difficult for particles larger than 30 nm. In order to provide maximum control on the NP structure along with synthetic straightforwardness, Navarro and co-workers used a 'one-pot' protocol, relying on a modified Turkevitch-Frens procedure. The method consist in the addition of sodium acetylacetonate (Na(acac) allowing the complexation and reduction of AuIII to AuI) immediately followed by the addition of sodium citrate (reduction of AuI to Au0). For this experiment, sodium acetylacetonate purity is really important. The concentration of sodium acetylacetonate directly impacts the nuclei numbers inducing a size evolution of the gold core up to 90 nm with a narrow size distribution. Practically, the tetrachloroaurate and sodium citrate concentration are fixed to 0.30 mM and 0.255 mM (ratio sodium citrate to gold: 0.85). In these conditions, and without sodium acetylacetonate, the obtained nanoparticles precipitate immediately. The concentration of sodium acetylacetonate can be tuned from 0.33 mM to 1.0 mM. The addition of the Na(acac) and sodium citrate to the yellow boiling gold ions solution results in a colourless one indicating the complexation and reduction of AuIII to AuI. Within a few minutes, the solution turns light blue-purple, featuring the formation of the gold nuclei. The small crystalline structures will then diffuse over the solution allowing the growth of the final spherical particles.

Sonolysis

Another method for the experimental generation of gold particles is by sonolysis. In one such process based on ultrasound, the reaction of an aqueous solution of $HAuCl_4$ with glucose, the reducing agents are hydroxyl radicals and sugar pyrolysis radicals (forming at the interfacial region between the collapsing cavities and the bulk water) and the morphology obtained is that of nanoribbons with width 30–50 nm and length of several micrometres. These ribbons are very flexible and can bend with angles larger than 90°. When glucose is replaced by cyclodextrin (a glucose oligomer), only spherical gold particles are obtained, suggesting that glucose is essential in directing the morphology toward a ribbon.

Block Copolymer-Mediated Method

An economical, environmentally benign and fast synthesis methodology for gold nanoparticles using block copolymer has been developed. In this synthesis methodology, block copolymer plays the dual role of a reducing agent as well as a stabilizing agent. The formation of gold nanoparticles comprises three main steps: reduction of gold salt ion by block copolymers in the solution and formation of gold clusters, adsorption of block copolymers on gold clusters and further reduction of gold salt ions on the surfaces of these gold clusters for the growth of gold particles in steps, and finally its stabilization by block copolymers. But this method usually has a limited-yield (nanoparticle concentration), which does not increase with the increase in the gold salt concentration. Recently, Ray et al. demonstrated that the presence of an additional reductant (trisodium citrate) in 1:1 molar ratio with gold salt enhances the yield by manyfold.

Electron Microscopy

Colloidal gold and various derivatives have long been among the most widely used labels for antigens in biological electron microscopy. Colloidal gold particles can be attached to many traditional biological probes such as antibodies, lectins, superantigens, glycans, nucleic acids, and receptors. Particles of different sizes are easily distinguishable in electron micrographs, allowing simultaneous multiple-labelling experiments.

Medical Research

Drug Carrier

Gold nanoparticles can be used to optimize the biodistribution of drugs to diseased organs, tissues or cells, in order to improve and target drug delivery. It is important to realise that the nanoparticle-mediated drug delivery is feasible only if the drug distribution is otherwise inadequate. These cases include drug targeting of difficult, unstable molecules (proteins, siRNA, DNA), delivery to the difficult sites (brain, retina, tumors, intracellular organelles) and drugs with serious side effects (e.g. anti-cancer agents). The performance of the nanoparticles depends on the size and surface functionalities in the particles. Also, the drug release and particle disintegration can vary depending on the system (eg. biodegradable polymers sensitive to pH). An optimal nanodrug delivery system ensures that the active drug is available at the site of action for the correct time and duration, and their concentration should be above the minimal effective concentration (MEC) and below the minimal toxic concentration (MTC).

Gold nanoparticles are being investigated as carriers for drugs such as Paclitaxel. The administration of hydrophobic drugs require molecular encapsulation and it is found that nanosized particles are particularly efficient in evading the reticuloendothelial system.

Gold nanoparticles are also used to circumvent multidrug resistance (MDR) mechanisms. Mechanisms of MDR include decreased uptake of drugs, reduced intracellular drug concentration by activation of the efflux transporters, modifications in cellular pathways by altering cell cycle checkpoints, increased metabolism of drugs, induced emergency response genes to impair apoptotic pathways and altered DNA repair mechanisms.

Tumor Detection

In cancer research, colloidal gold can be used to target tumors and provide detection using SERS (Surface Enhanced Raman Spectroscopy) *in vivo*. These gold nanoparticles are surrounded with Raman reporters, which provide light emission that is over 200 times brighter than quantum dots. It was found that the Raman reporters were stabilized when the nanoparticles were encapsulated with a thiol-modified polyethylene glycol coat. This allows for compatibility and circulation *in vivo*. To specifically target tumor cells, the pegylated gold particles are conjugated with an antibody (or an antibody fragment such as scFv), against, e.g. Epidermal growth factor receptor, which is sometimes overexpressed in cells of certain cancer types. Using SERS, these pegylated gold nanoparticles can then detect the location of the tumor.

Gold nanoparticles accumulate in tumors, due to the leakiness of tumor vasculature, and can be used as contrast agents for enhanced imaging in a time-resolved optical tomography system using short-pulse lasers for skin cancer detection in mouse model. It is found that intravenously administrated spherical gold nanoparticles broadened the temporal profile of reflected optical signals and enhanced the contrast between surrounding normal tissue and tumors.

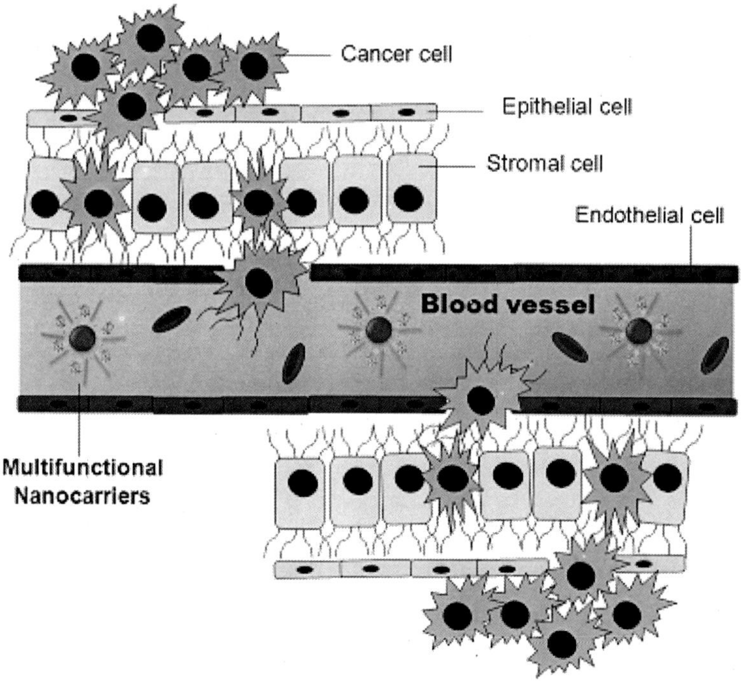

Figure: *Tumor targeting via multifunctional nanocarriers. Cancer cells reduce adhesion to neighbouring cells and migrate into the vasculature-rich stroma. Once at the vasculature, cells can freely enter the bloodstream. Once the tumor is directly connected to the main blood circulation system, multifunctional nanocarriers can interact directly with cancer cells and effectively target tumors.*

Therefore, gold nanoparticles have the potential to join numerous therapeutic functions into a single platform, by targeting specific tumor cells, tissues and organs. Actually, Conde et al reported the evaluation of the inflammatory response and therapeutic siRNA silencing via RGD-nanoparticles in a lung cancer mouse model. This study reported the use of siRNA/RGD gold nanoparticles capable of targeting tumor cells in two lung cancer xenograft mouse models, resulting in successful and significant *c-Myc* oncogene downregulation followed by tumor growth inhibition and prolonged survival of the animals. This delivery system can achieve translocation of siRNA duplexes directly into the tumour cell cytoplasm and accomplish successful silencing of an oncogene expression. Actually, RGD/siRNA-AuNPs can target preferentially and be taken up by tumor cells via integrin $\alpha v\beta 3$-receptor-mediated endocytosis with no cytotoxicity, showing that can accumulate in tumor tissues overexpressing $\alpha v\beta 3$ integrins and selectively delivered *c-Myc* siRNA to suppress tumor growth and angiogenesis.

Gene Therapy

Gene therapy is receiving increasing attention and, in particular, small-interference RNA (siRNA) shows importance in novel molecular approaches in the knockdown of specific gene expression in cancerous cells. The major obstacle to clinical application is the uncertainty about how to deliver therapeutic siRNAs with maximal therapeutic impact. Gold nanoparticles have shown potential as intracellular delivery vehicles for siRNA oligonucleotides with maximal therapeutic impact.

Recently, Conde et al. provided evidence of *in vitro* and *in vivo* RNAi triggering via the synthesis of a library of novel multifunctional gold nanoparticles, using a hierarchical approach including three biological systems of increasing complexity: *in vitro* cultured human cells, *in vivo* freshwater polyp (*Hydra vulgaris*), and *in vivo* mice models. The authors developed effective conjugation strategies to combine, in a highly controlled way, specific biomolecules to the surface of gold nanoparticles such as: (a) biofunctional spacers: Poly(ethylene glycol) (PEG) spacers used to increase solubility and biocompatibility; (b) cell penetrating peptides such as TAT and RGD peptides: A novel class of membrane translocating agents named cell penetrating peptides (CPPs) that exploit more than one mechanism of endocytosis to overcome the lipophilic barrier of the cellular membranes and deliver large molecules and even small particles inside the cell for their biological actions; and (c) siRNA complementary to a master regulator gene, the protooncogene c-myc, were bond covalently (thiol-siRNA) and ionically (naked/unmodified siRNA) to gold nanoparticles.

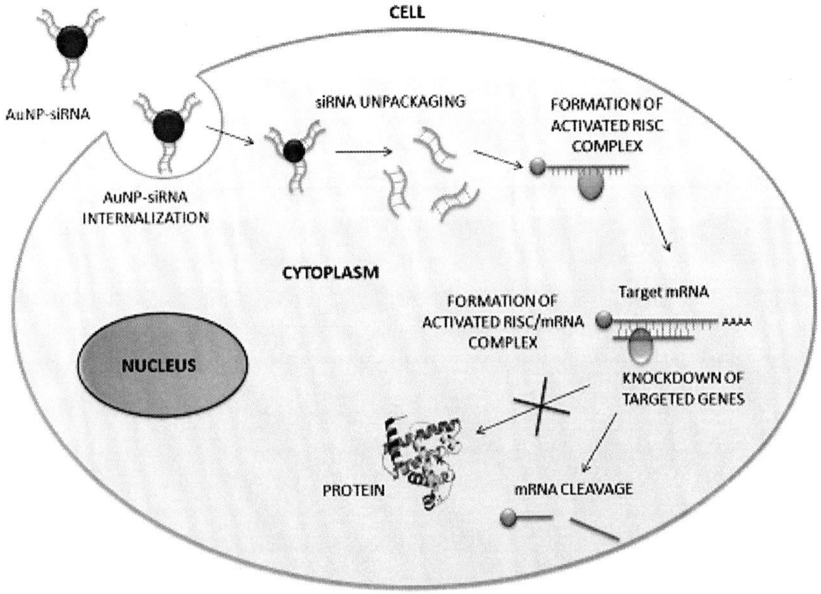

Figure: *Regulation of gene expression via siRNA-gold nanoparticles. Engineered nanoparticles modified with siRNA can represent a delivery system of siRNA and a useful tool to block gene function and for sequence-specific post-transcriptional gene silencing. siRNA-modified nanocarriers can enter cells and subsequently unwound siRNA strands and assembled into an effector complex, RNA Induced Silencing Complex (RISC), which can direct RNA cleavage, mediate translational repression or induce chromatin modification. The antisense strand then binds to its complementary/target mRNA (activated RISC/mRNA complex). The catalytic RISC recognises mRNAs containing perfect or near-perfect complementary sequence to the guide siRNA and cleaves the mRNAs at a site precisely 10 nucleotides from the 5'-end of the guide strand. Finally, mRNA degradation is achieved by endo- and exonucleases, resulting in knockdown of the expression of the corresponding genes.*

Gold nanoparticles have also shown potential as intracellular delivery vehicles for antisense oligonucleotides (ssDNA,dsDNA) by providing protection against intracellular nucleases and ease of functionalization for selective targeting. Recently, Conde et al developed a new theranostic system capable of intersecting all RNA pathways: from gene specific downregulation to silencing the silencers, i.e. siRNA and miRNA pathways. The authors reported the development gold nanoparticles functionalized with a fluorophore labelled hairpin-DNA, i.e. gold nanobeacons, capable of efficiently silencing single gene expression, exogenous siRNA and endogenous miRNAs while yielding a quantifiable fluorescence signal directly proportional to the level of silencing. This method describes a gold nanoparticle-based nanobeacon as an innovative theranostic approach for detection and inhibition of sequence-specific DNA and RNA for *in vitro* and *ex vivo* applications. Under hairpin configuration, proximity to gold nanoparticles leads to fluorescence quenching; hybridization to a complementary target restores fluorescence emission due to the gold nanobeacons' conformational reorganisation that causes the fluorophore and the gold nanoparticle to part from each other. This concept can easily be extended and adapted to assist the in vitro evaluation of silencing potential of a given sequence to be later used for *ex vivo* gene silencing and RNAi approaches, with the ability to monitor real-time gene delivery action.

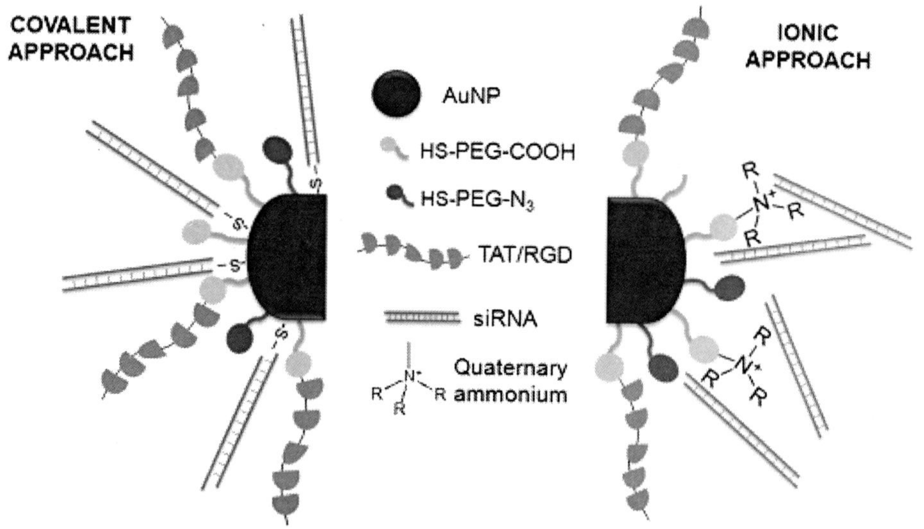

Figure: *Multifunctional siRNA-gold nanoparticles with several biomolecules: PEG, cell penetration and cell adhesion peptides and siRNA. Two different approaches were employed to conjugate the siRNA to the gold nanoparticle: (1) Covalent approach: use of thiolated siRNA for gold-thiol binding to the nanoparticle; (2) Ionic approach: interaction of the negatively charged siRNA to the modified surface of the AuNP through ionic interactions.*

Photothermal Agents

Gold nanorods are being investigated as photothermal agents for in-vivo applications. Gold nanorods are rod-shaped gold nanoparticles whose aspect ratios tune the surface plasmon resonance (SPR) band from the visible to near-infrared wavelength. The total extinction of light at the SPR is made up of both absorption and scattering. For the smaller axial diameter nanorods (~10 nm), absorption dominates, whereas for the larger axial diameter nanorods (>35 nm) scattering can dominate. As a consequence, for in-vivo applications, small diameter gold nanorods are being used as photothermal converters of near-infrared light due to their high absorption cross-sections. Since near-infrared light transmits readily through human skin and tissue, these nanorods can be used as ablation components for cancer, and other targets. When coated with polymers, gold nanorods have been known to circulate in-vivo for greater than 15 hours half-life. Apart from rod-like gold nanoparticles, also spherical colloidal gold nanoparticles are recently used as markers in combination with photothermal single particle microscopy.

Radiotherapy Dose Enhancer

Following work by Hainfield et al. there has been considerable interest in the use of gold and other heavy-atom containing nanoparticles to enhance the dose delivered to tumors. Since the gold nanoparticles are taken up by the tumors more than the nearby healthy tissue, the dose is selectively enhanced. The biological effectiveness of this type of therapy seems to be due to the local deposition of the radiation dose near the nanoparticles. This mechanism is the same as occurs in heavy ion therapy.

Gold Nanobeacons

Gold nanobeacons are gold nanoparticles functionalized with a fluorophore-labelled hairpin-DNA, i.e. gold-nanobeacon to follow RNA synthesis in real time in bulky solutions and for antisense DNA and RNA interference (RNAi), from gene specific silencing to silence-the-silencers. Under hairpin configuration, proximity to gold nanoparticles leads to fluorescence quenching; hybridization to a complementary target restores fluorescence emission due to the gold nanobeacons' conformational reorganisation that causes the fluorophore and the gold nanoparticle to part from each other.

Gold Nanobeacons for Theranostics

Gold nanoparticle-based molecular beacon have shown the capability to specifically detect DNA target sequences, with better discriminating power for single-mismatch than regular molecular beacons. Nevertheless, the systems described so far rely on the detection of nucleic acids in bulk solution and can only be used in vitro applications without translation to cell/tissue/organism testing.

Recently, Conde et al. developed a new theranostic system capable of intersecting all RNA pathways: from gene specific downregulation to silencing the silencers, i.e. siRNA and miRNA pathways. The authors reported the development of new nanomaterials, i.e. gold nanoparticles functionalized with a fluorophore labelled hairpin-DNA - Gold nanobeacons - capable of efficiently silencing single gene expression, exogenous siRNA and endogenous miRNAs while yielding a quantifiable fluorescence signal directly proportional to the level of silencing.

This method describes a gold nanoparticle-based nanobeacon as an innovative theranostic approach for detection and inhibition of sequence-specific DNA and RNA for *in vitro* and *ex vivo* applications. This concept can easily be extended and adapted to assist the in vitro evaluation of silencing potential of a given sequence to be later used for ex vivo gene silencing and RNAi approaches, with the ability to monitor real-time gene delivery action.

The mechanism proposed in the gold-nanobeacon tool has got several clear advantages compared with traditional methods. Firstly, naked/unmodified oligonucleotides show extremely short half-lives inside cellular environment, feeble protection against RNases and other nucleases, poor chemical stability, and common dissociation from vector. In fact, the major obstacle to clinical application is the uncertainty about how to deliver therapeutic DNAs (antisense oligonucleotides) and RNAs (e.g., microRNA and/or siRNA) with maximal therapeutic impact due to systemic loss of cargo from traditional vectors. This method offers an unprecedented opportunity to overcome these problems as these nanoconjugates can readily interact with biomolecules on both the surface of cells and inside cells for longer periods of time, due to their small size and protective environment for DNA/RNA oligonucleotides provided by the metal nanoparticle core. The gold-nanobeacons are also highly soluble, homogenous and stable and are not prone to aggregation.

In addition, these nanoparticles are thermodynamically stable and can remain inside cells for long periods of time and at low concentrations. Secondly, gold-nanobeacons are capable of efficiently silencing single gene expression, exogenous siRNA and endogenous miRNAs. Thirdly, it allows real-time detection of the beacon's signal while yielding a quantifiable fluorescence directly proportional to the level of gene silencing. This can be used to track the silencing events inside the cell as they occur. All this is achieved in a simple approach that can be straightforwardly adapted and tailored to any specific target. Finally, a significant attribute of these gold-nanobeacons is the ability to attain similar levels of inhibition of gene expression with lower amounts than those of free oligonucleotides without increasing cell death. This extraordinary efficiency occurs probably due to the large payload capacity of the NPs and the longer half-life when inside the cells. The gold-nanobeacons may represent in the near future an economically viable and commercial-scale production for a cell and cell-free system.

Contrary to conventional gene delivery chemistry, which is often associated to systemic toxicity and adverse effects as well as lack of specificity and lower product life-cycle, this method represents a safe, efficient, specific and non-pathogenic vehicle for gene delivery and cell tracking system – a theranostic tool.

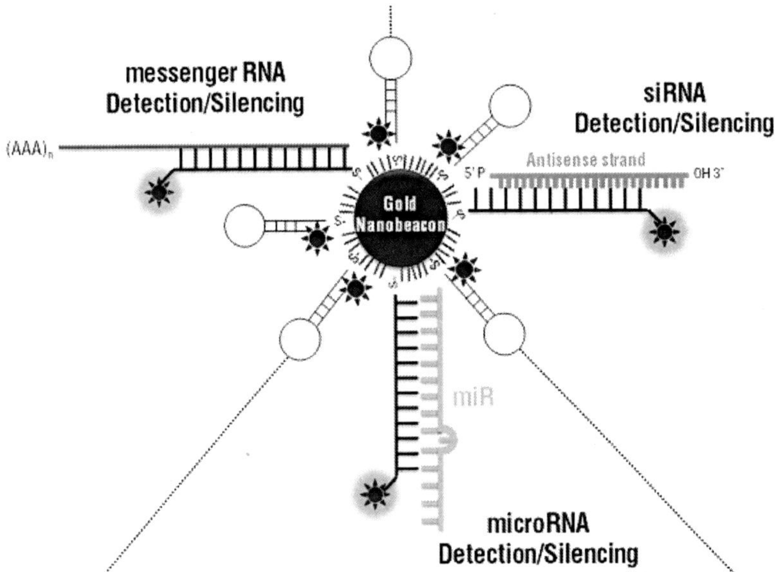

Universal gene silencing tool based on gold nanoparticles functionalized with a fluorophore labelled hairpin-DNA – Gold nanobeacon (Au-nanobeacon) - that effectively detects and silences the specific target (siRNA, miRNA or gene specific mRNA) while simultaneously signalling its action via fluorescence emission in cancer cells.

Colloidal Silver

Colloidal silver (a colloid consisting of silver particles suspended in liquid) and formulations containing silver salts were used by physicians in the early 20th century, but their use was largely discontinued in the 1940s following the development of safer and effective modern antibiotics. Since the 1990s, colloidal silver has again been marketed as an alternative medicine, often with extensive "cure-all" claims. Colloidal silver products remain available in many countries as dietary supplements and homeopathic remedies, although they are not effective in treating any known condition and carry the risk of both permanent cosmetic side effects such as argyria and more serious ones such as allergic reactions, and interactions with prescription medications.

Nanoengineering

Nanoengineering is the practice of engineering on the nanoscale. It derives its name from the nanometre, a unit of measurement equalling one billionth of a metre. Nanoengineering is largely a synonym for nanotechnology, but emphasizes the engineering rather than the pure science aspects of the field.

Degree Programs

The first nanoengineering program in the world was started at the University of Toronto within the Engineering Science program as one of the Options of study in the final years. In 2003, the Lund Institute of Technology started a program in Nanoengineering. In 2004, the College of Nanoscale Science and Engineering(CNSE) was established on the campus of the University at Albany. In 2005, the University of Waterloo established a unique program which offers a full degree in Nanotechnology Engineering. Louisiana Tech University started the first program in the U.S. in 2005. In 2006 the University of Duisburg-Essen started a Bachelor and a Master program NanoEngineering. The University of California, San Diego followed shortly thereafter in 2007 with its own department of Nanoengineering. In 2009, the University of Toronto began offering all Options of study in Engineering Science as degrees, bringing the second nanoengineering degree to Canada. DTU Nanotech - the Department of Micro- and Nanotechnology - is a department at the Technical University of Denmark established in 1990.

Techniques

- Scanning tunneling microscope (STM) - Can be used to both image, and to manipulate structures as small as a single atom.
- Molecular self-assembly - Arbitrary sequences of DNA can now be synthesized cheaply in bulk, and used to create custom proteins or regular patterns of amino acids. Similarly, DNA strands can bind to other DNA strands, allowing simple structures to be created.

Nanobiotechnology

Bionanotechnology, nanobiotechnology, and nanobiology are terms that refer to the intersection of nanotechnology and biology. Given that the subject is one that has only emerged very recently, bionanotechnology and nanobiotechnology serve as blanket terms for various related technologies.

This discipline helps to indicate the merger of biological research with various fields of nanotechnology. Concepts that are enhanced through nanobiology include: nanodevices, nanoparticles, and nanoscale phenomena that occurs within the discipline of nanotechnology. This technical approach to biology allows scientists to imagine and create systems that can be used for biological research. Biologically inspired nanotechnology uses biological systems as the inspirations for technologies not yet created. However, as with nanotechnology and biotechnology, bionanotechnology does have many potential ethical issues associated with it.

The most important objectives that are frequently found in nanobiology involve applying nanotools to relevant medical/biological problems and refining these applications. Developing new tools, such as peptoid nanosheets, for medical and biological purposes is another primary objective in nanotechnology. New nanotools are often made by refining the applications of the nanotools that are already being used. The imaging of native biomolecules, biological membranes, and tissues is also a major topic for the nanobiology researchers. Other topics concerning nanobiology include the use of cantilever array sensors and the application of nanophotonics for manipulating molecular processes in living cells.

Recently, the use of microorganisms to synthesize functional nanoparticles has been of great interest. Microorganisms can change the oxidation state of metals. These microbial processes have opened up new opportunities for us to explore novel applications, for example, the biosynthesis of metal nanomaterials. In contrast to chemical and physical methods, microbial processes for synthesizing nanomaterials can be achieved in aqueous phase under gentle and environmentally benign conditions. This approach has become an attractive focus in current green bionanotechnology research towards sustainable development.

Terminology

The terms are often used interchangeably. When a distinction is intended, though, it is based on whether the focus is on applying biological ideas or on studying biology with nanotechnology. Bionanotechnology generally refers to the study of how the goals of nanotechnology can be guided by studying how biological "machines" work and adapting these biological motifs into improving existing nanotechnologies or creating new ones. Nanobiotechnology, on the other hand, refers to the ways that nanotechnology is used to create devices to study biological systems.

In other words, nanobiotechnology is essentially miniaturized biotechnology, whereas bionanotechnology is a specific application of nanotechnology. For example, DNA nanotechnology or cellular engineering would be classified as bionanotechnology because they involve working with biomolecules on the nanoscale. Conversely, many new medical technologies involving nanoparticles as delivery systems or as sensors would be examples of nanobiotechnology since they involve using nanotechnology to advance the goals of biology.

The definitions enumerated above will be utilised whenever a distinction between nanobio and bionano is made in this article. However, given the overlapping usage of the terms in modern parlance, individual technologies may need to be evaluated to determine which term is more fitting. As such, they are best discussed in parallel.

Concepts

Most of the scientific concepts in bionanotechnology are derived from other fields. Biochemical principles that are used to understand the material properties of biological systems are central in bionanotechnology because those same principles are to be used to create new technologies. Material properties and applications studied in bionanoscience include mechanical properties (e.g. deformation, adhesion, failure), electrical/electronic (e.g. electromechanical stimulation, capacitors, energy storage/batteries), optical (e.g. absorption, luminescence, photochemistry), thermal (e.g. thermomutability, thermal management), biological (e.g. how cells interact with nanomaterials, molecular flaws/defects, biosensing, biological mechanisms s.a. mechanosensing), nanoscience of disease (e.g. genetic disease, cancer, organ/tissue failure), as well as computing (e.g. DNA computing). The impact of bionanoscience, achieved through structural and mechanistic analyses of biological processes at nanoscale, is their translation into synthetic and technological applications through nanotechnology.

Nano-biotechnology takes most of its fundamentals from nanotechnology. Most of the devices designed for nano-biotechnological use are directly based on other existing nanotechnologies. Nano-biotechnology is often used to describe the overlapping multidisciplinary activities associated with biosensors, particularly where photonics, chemistry, biology, biophysics, nano-medicine, and engineering converge. Measurement in biology using wave guide techniques, such as dual polarization interferometry, are another example.

Applications

Applications of bionanotechnology are extremely widespread. Insofar as the distinction holds, nanobiotechnology is much more commonplace in that it simply provides more tools for the study of biology. Bionanotechnology, on the other hand, promises to recreate biological mechanisms and pathways in a form that is useful in other ways.

Nanobiotechnology (sometimes referred to as nanobiology) is best described as helping modern medicine progress from treating symptoms to generating cures and regenerating biological tissues. Three American patients have received whole cultured bladders with the help of doctors who use nanobiology techniques in their practice. Also, it has been demonstrated in animal studies that a uterus can be grown outside the body and then placed in the body in order to produce a baby. Stem cell treatments have been used to fix diseases that are found in the human heart and are in clinical trials in the United States. There is also funding for research into allowing people to have new limbs without having to resort to prosthesis. Artificial proteins might also become available to manufacture without the need for harsh chemicals and expensive machines. It has even been surmised that by the year 2055, computers may be made out of biochemicals and organic salts.

Another example of current nanobiotechnological research involves nanospheres coated with fluorescent polymers. Researchers are seeking to design polymers whose fluorescence is quenched when they encounter specific molecules. Different polymers would detect different metabolites. The polymer-coated spheres could become part of new biological assays, and the technology might someday lead to particles which could be introduced into the human body to track down metabolites associated with tumors and other health problems. Another example, from a different perspective, would be evaluation and therapy at the nanoscopic level, i.e. the treatment of Nanobacteria (25-200 nm sized) as is done by NanoBiotech Pharma.

While nanobiology is in its infancy, there are a lot of promising methods that will rely on nanobiology in the future. Biological systems are inherently nano in scale; nanoscience must merge with biology in order to deliver biomacromolecules and molecular machines that are similar to nature. Controlling and mimicking the devices and processes that are constructed from molecules is a tremendous challenge to face the converging disciplines of nanotechnology. All living things, including humans, can be considered to be nanofoundries. Natural evolution has optimized the "natural" form of nanobiology over millions of years. In the 21st century, humans have developed the technology to artificially tap into nanobiology. This process is best described as "organic merging with synthetic." Colonies of live neurons can live together on a biochip device; according to research from Dr. Gunther Gross at the University of North Texas. Self-assembling nanotubes have the ability to be used as a structural system. They would be composed together with rhodopsins; which would facilitate the optical computing process and help with the storage of biological materials. DNA (as the software for all living things) can be used as a structural proteomic system - a logical component for molecular computing. Ned Seeman - a researcher at New York University - along with other researchers are currently researching concepts that are similar to each other.

Bionanotechnology

DNA nanotechnology is one important example of bionanotechnology. The utilisation of the inherent properties of nucleic acids like DNA to create useful materials is a promising area of modern research. Another important area of research involves taking advantage of membrane properties to generate synthetic membranes. Protein folding studies provide a third important avenue of research, but one that has been largely inhibited by our inability to predict protein folding with a sufficiently high degree of accuracy. Given the myriad uses that biological systems have for proteins, though, research into understanding protein folding is of high importance and could prove fruitful for bionanotechnology in the future.

Lipid nanotechnology is another major area of research in bionanotechnology, where physico-chemical properties of lipids such as their antifouling and self-assembly is exploited to build nanodevices with applications in medicine and engineering.

Tools

This field relies on a variety of research methods, including experimental tools (e.g. imaging, characterization via AFM/optical tweezers etc.), x-ray diffraction based tools, synthesis via self-assembly, characterization of self-assembly (using e.g. dual polarization interferometry, recombinant DNA methods, etc.), theory (e.g. statistical mechanics, nanomechanics, etc.), as well as computational approaches (bottom-up multi-scale simulation, supercomputing).

Nanofluidics

Nanofluidics is the study of the behaviour, manipulation, and control of fluids that are confined to structures of nanometre (typically 1-100 nm) characteristic dimensions (1 nm = 10^{-9} m). Fluids confined in these structures exhibit physical behaviours not observed in larger structures, such as those of micrometre dimensions and above, because the characteristic physical scaling lengths of the fluid, (*e.g.* Debye length, hydrodynamic radius) very closely coincide with the dimensions of the nanostructure itself.

When structures approach the size regime corresponding to molecular scaling lengths, new physical constraints are placed on the behaviour of the fluid. For example, these physical constraints induce regions of the fluid to exhibit new properties not observed in bulk, *e.g.* vastly increased viscosity near the pore wall; they may effect changes in thermodynamic properties and may also alter the chemical reactivity of species at the fluid-solid interface. A particularly relevant and useful example is displayed by electrolyte solutions confined in nanopores that contain surface charges, *i.e.* at electrified interfaces, as shown in the nanocapillary array membrane (NCAM) in the accompanying figure.

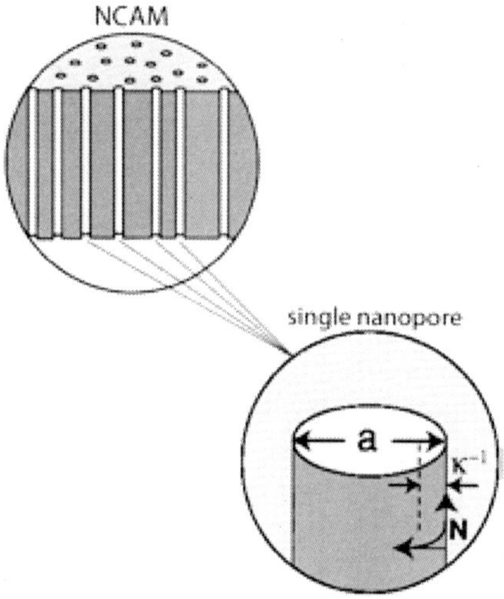

Figure: *Schematic diagram of one particular realisation of nanofluidics in a nanocapillary array membrane, or NCAM. The NCAM is composed of a large number of parallel nanocapillaries, each of which have a pore radius,* a/2, *which is approximately the same size as the Debye length,* k^{-1}. *The electrical double layer is characterized by a counter-ion distribution,* N, *which is largest at the pore wall and decays toward the centre of the pore.*

All electrified interfaces induce an organised charge distribution near the surface known as the electrical double layer. In pores of nanometre dimensions the electrical double layer may completely span the width of the nanopore, resulting in dramatic changes in the composition of the fluid and the related properties of fluid motion in the structure. For example, the drastically enhanced surface-to-volume ratio of the pore results in a preponderance of counter-ions (*i.e.* ions charged oppositely to the static wall charges) over co-ions (possessing the same sign as the wall charges), in many cases to the near-complete exclusion of co-ions, such that only one ionic species exists in the pore. This can be used for manipulation of species with selective polarity along the pore length to achieve unusual fluidic manipulation schemes not possible in micrometre and larger structures.

Theory

In 1965, Rice and Whitehead published the seminal contribution to the theory of the transport of electrolyte solutions in long (ideally infinite) nanometre-diameter capillaries. Briefly, the potential, φ, at a radial distance, *r*, is given by the Poisson-Boltzmann equation,

$$\frac{1}{r}\frac{d}{dr}\left(r\frac{d\phi}{dr}\right)=\kappa^2\phi$$

where κ is the inverse Debye length,

$$\kappa=\sqrt{\frac{8\pi ne^2}{\epsilon kT}}$$

determined by the ion number density, *n*, the dielectric constant, ε, the Boltzmann constant, *k*, and the temperature, *T*. Knowing the potential, *φ(r)*, the charge density can then be recovered from the Poisson equation, whose solution may be expressed as a modified Bessel function of the first kind, I_0, and scaled to the capillary radius, *a*. An equation of motion under combined pressure and electrically-driven flow can then be written,

$$\frac{1}{r}\frac{d}{dr}\left(r\frac{dv_z}{dr}\right)=\frac{1}{\eta}\frac{dp}{dz}-\frac{F_z}{\eta}$$

where η is the viscosity, dp/dz is the pressure gradient, and F_z is the body force driven by the action of the applied electric field, E_z, on the net charge density in the double layer. When there is no applied pressure, the radial distribution of the velocity is given by,

$$v_z(r)=\frac{\epsilon\phi_0}{4\pi\eta}E_z\left[1-\frac{I_0(\kappa r)}{I_0(\kappa a)}\right]$$

From the equation above, it follows that fluid flow in nanocapillaries is governed by the κ*a* product, that is, the relative sizes of the Debye length and the pore radius. By adjusting these two parameters and the surface charge density of the nanopores, fluid flow can be manipulated as desired.

Fabrication

Nanostructures can be fabricated as single cylindrical channels, nanoslits, or nanochannel arrays from materials such as silicon, glass, polymers (e.g. PMMA, PDMS, PCTE) and synthetic vesicles. Standard photolithography, bulk or surface micromachining, replication techniques (embossing, printing, casting and injection molding), and nuclear track or chemical etching, are commonly used to fabricate structures which exhibit characteristic nanofluidic behaviour.

Applications

Because of the small size of the fluidic conduits, nanofluidic structures are naturally applied in situations demanding that samples be handled in exceedingly small quantities, including Coulter counting, analytical separations and determinations of biomolecules, such as proteins and DNA, and facile handling of mass-limited samples. One of the more promising areas of nanofluidics is its potential for integration into microfluidic systems, i.e. MicroTotal Analytical Systems or Lab-on-a-chip structures.

For instance, NCAMs, when incorporated into microfluidic devices, can reproducibly perform digital switching, allowing transfer of fluid from one microfluidic channel to another, selectivity separate and transfer analytes by size and mass, mix reactants efficiently, and separate fluids with disparate characteristics. In addition, there is a natural analogy between the fluid handling capabilities of nanofluidic structures and the ability of electronic components to control the flow of electrons and holes.

This analogy has been used to realise active electronic functions such as rectification and field-effect and bipolar transistor action with ionic currents. Application of nanofluidics is also to Nano-optics for producing tuneable microlens array.

Nanofluidics have had a significant impact in biotechnology, medicine and clinical diagnostics with the development of lab-on-a-chip devices for PCR and related techniques.

Because the science of nanofluidics is still in its infancy, we can expect rapid development of new applications in the coming years.

Challenges

There are a variety of challenges associated with the flow of liquids through carbon nanotubes and nanopipes. A common occurrence is channel blocking due to large macromolecules in the liquid. Also, any insoluble debris in the liquid can easily clog the tube. A solution for this researchers are hoping to find is a low friction coating or channel materials that help reduce the blocking of the tubes. Also, due to the large size of polymers, including biologically relevant molecules such as DNA often fold *in vivo*. This causes blockage because typical DNA molecules from a virus have lengths of approx. 100-200 kilobases and will form a random coil of the radius some 700 nm in aqueous solution at 20%. This is also several times greater than the pore diameter of even large carbon pipes and 2 orders of magnitude the diameter of a single walled carbon nanotube.

Nanofluidic Circuitry

Nanofluidic circuitry is a nanotechnology aiming for control of fluids in nanometre scale. Due to the effect of an electrical double layer within the fluid channel, the behaviour of nanofluid is observed to be significantly different compared with its microfluidic counterparts. Its typical characteristic dimensions fall within the range of 1–100 nm. At least one dimension of the structure is in nanoscopic scale. Phenomena of fluids in nano-scale structure are discovered to be of different properties in electrochemistry and fluid dynamics.

Background

With the development of microfabrication and nanotechnology, the study of microfluidics and nanofluidics is drawing more attention. Research on microfluidic found its advantages in DNA analysis, lab-on-a-chip, and micro-TAS. Devices in a microfluidic system include channels, valves, mixers, and pumps. Integration of these microfluidic devices enables sorting, transporting, and mixing of substances within fluids. However, the failure of moving parts in these systems is usually the critical issue and the main drawback. Mechanisms to control flow without using mechanical parts are always desired for reliability and lifetime.

In 1997, Chang Wei and his colleague discovered that ion rectification occurs at the tip of a nano-sized pipe. They observed that the surface charge at the wall of a nano-pipet induced a non-neutral electrical potential within the orifice. The electrical potential then modifies the concentration of ion species, resulting in an asymmetric current-voltage characteristic for the current through the pipet.

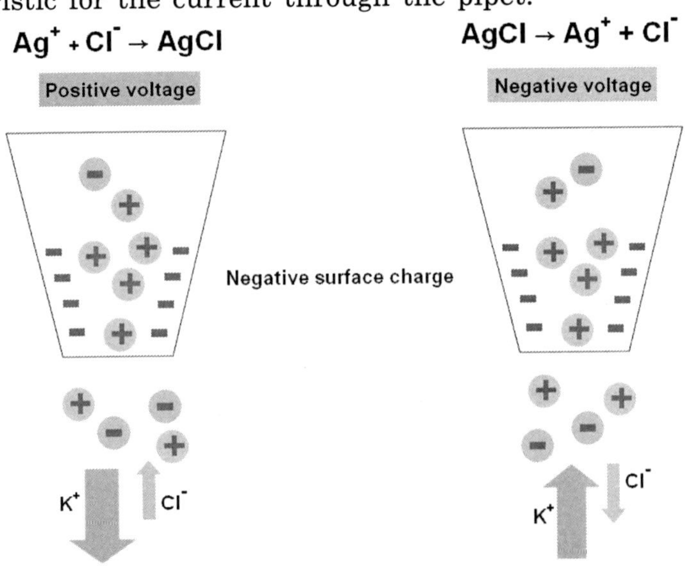

Transport of ions in the electrolyte can be adjusted by tuning the pH value in a dilute ionic solution, or by introducing an external electrical potential to change the surface charge density of the wall. As an analogy to semiconductor devices, the mechanism to control charge carriers transport in electronic devices was established in the area of nanofluidics.

In nanofluidics, the active control of ion transport is realised using nano-scale channels or pores.

Research efforts on micro-scaled fluidic systems started to focus on the rectifying phenomena, which can be seen only in nano-scaled systems. In 2006, Professor Majumdar and Professor Yang in University of California, Berkeley built the first "nanofluidic" transistor. The transistor can be turn on or off by an external electrical signal, allowing the control of ionic fluids in a nano-scaled channel. Their work implies a possibility to create a nanofluidic circuitry with logic functions.

The main researchers in the area of nanofluidic devices include Arun Majumdar and Peidong Yang in University of California - Berkeley, Harold Craighead and Brian Kirbyat Cornell University, Juan Santiago at Stanford University, Albert van den Berg in University of Twente, Zuzanna Siwy in University of California - Irvine, and Mark Shannon in University of Illinois - Urbana-Champaign.

Basic Principles

For electrolyte solution in a channel with a macro- or micro-scaled radius, surface charges at the wall attract counterions and repel co-ions due to electrostatic force. Therefore, an electrical double layer exists between the wall of channel and the solution. The dimension of the electrical double layer is determined by the Debye length in this system, which is typically much smaller than the channel radius. Most of the solution in the channel is electrically neutral due to the shielding effect of the electrical double layer.

In a nanochannel, however, the solution is charged when the dimension of channel radius is smaller than the Debye length. Therefore, it is possible to manipulate the flow of ions inside the nanochannel by introducing surface charges on the wall or by applying an external electrical potential.

Ionic concentration of solution has an important effect on the ion transport. Because a higher concentration leads to a shorter Debye length for the electrical double layer at the channel wall. Its rectifying effect decreases with the increase of ionic concentration. On the other hand, the rectification of ion can be improved by having a dilute solution.

Ion Transport

To analyze the transport of ions in the channel, behaviours of system in electrochemistry as well as fluid mechanics need to be considered. The Poisson–Nernst–Planck (PNP) equations are utilised to describe ionic current flowing through a channel, and the Navier–Stokes (NS) equations are used to represent the fluid dynamics in the channel.

The PNP equations consist of the Poisson equation:

$$\nabla^2\phi = -\frac{1}{\varepsilon_0\varepsilon}\sum_a z_a e n_a$$

and the Nernst–Planck equations, which gives the particle flux of ion species α due to a concentration gradient and electric potential gradient:

$$J_a = -D_a(\nabla n_a + \frac{z_a e n_a}{kT}\nabla\phi)$$

where ϕ is the electrostatic potential, e is the unit charge of electron, ε_0 is the permittivity in vacuum, and ε is the dielectric constant of solution; D_a, n_a and z_a are the diffusivity, the number density of ions, and the valence of ion species a.

The solution in steady-state satisfies the continuity equation. To describe fluid velocity field in the channel, using Navier–Stokes equations:

$$\nabla\cdot(n_a \mathrm{u} + J_a) = 0$$

$$\nabla\cdot \mathrm{u} = 0$$

$$\mathrm{u}\cdot\nabla \mathrm{u} = \frac{1}{\rho}[-\nabla p + \mu\nabla^2 \mathrm{u} - (\sum_a z_a e n_a)\nabla\phi]$$

where p,u , μ, and ρ are pressure, velocity vector, viscosity, and density of fluid, respectively. The equations above are usually solved with numerical algorithm to determine the velocity, pressure, electric potential, and ionic concentration in the fluid, as well as the electric current flow through the channel.

Ionic Selectivity

Ionic selectivity is defined to evaluate the performance of a nano-channel for ionic flow control. Ionic selectivity is the ratio of the difference in currents of majority and minority carriers to the total current carried by both positive and negative ions, *I*. For a nanochannel with perfect control over cation and anion, the selectivity is unity. For a nanochannel without ionic flow control, the selectivity is zero.

$$S = \frac{I^+ - I^-}{I^+ + I^-}$$

Nanofluidic Logic Devices

- Transport is proportional to applied bias (resistor)
- Transport can be made to move in one direction (diode)
- Control of gain is possible by introduction of third pole (transistor)
- Control of the forward/reverse direction by asymmetric gates (Field-effect reconfigurable diode)

Diodes

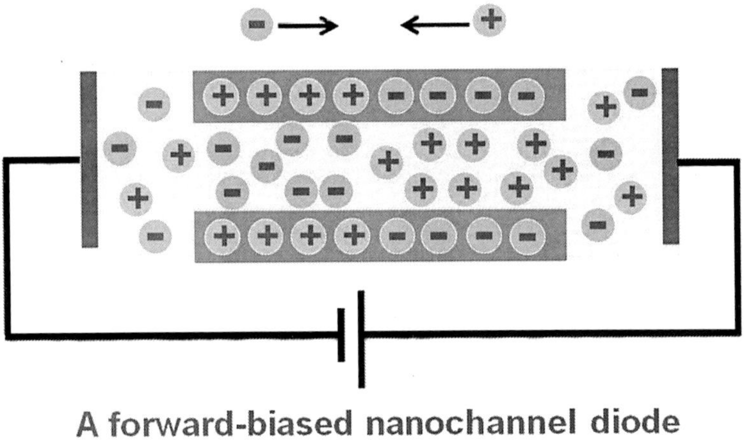

Figure: *A forward-biased nanofluidic diode*

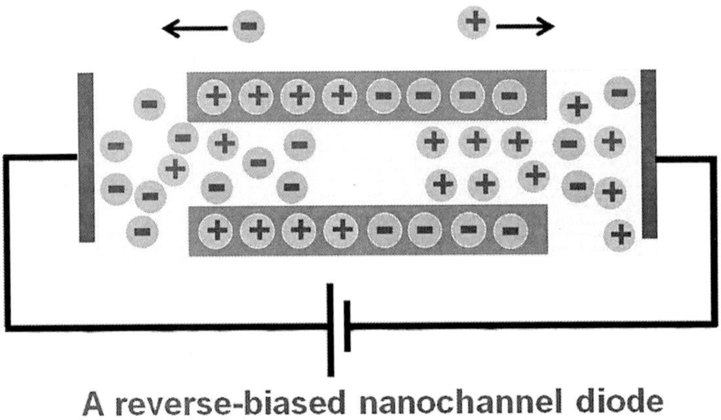

Figure: *A reverse-biased nanofluidic diode*

Nanofluidic diodes are utilised for rectification of ionic transport. A diode in electronic circuits limits the flow of electric current to one direction. A nanofluidic diode has the same function to restrict the ionic flow in one direction. A nanofluidic diode is a channel with its radius dimension of several nanometres. The inner surface of the channel is coated with surface charges. Current rectification can occur when the surface charges at the wall are of the same sign. It is also observed that, when a half of the channel is coated with opposite sign or electrically neutral, the rectification will be enhanced.

When the wall of the channel is coated with positive charges, the negative charged ions in the electrolyte will be attracted and accumulated within the channel. In this case, the flow of positive charges passing through the channel is not favourable, resulting in a decrease in ionic current. Therefore, the ionic current becomes asymmetric if the biasing voltage is reversed.

Field-Effect Transistors

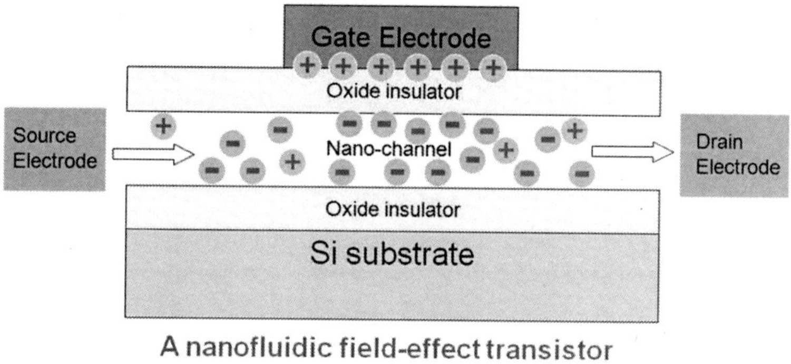

Figure: *A nanofluidic field-effect transistor*

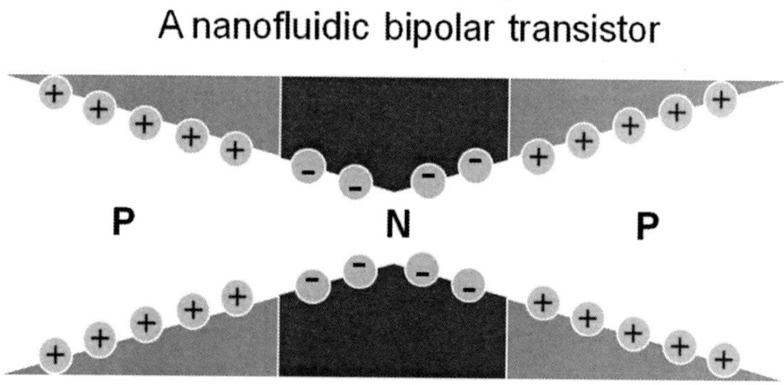

Figure: *A nanofluidic bipolar transistor*

By applying an additional electrode on a nanochannel as the gate electrode, it is possible to adjust the electrical potential inside the channel. A nanofluidic field-effect transistor can be made of silica nanotubes with an oxide as the dielectric material between the metal gate electrode and the channel. The tuning of the ionic current, therefore, can be achieved by changing the voltage applied on the gate. The gate bias and the source-drain bias are applied to adjust the cation and anion concentration within the nanochannel, therefore tuning the ionic current flowing through it.

This concept is an analogy to the structure of a metal-oxide semiconductor field-effect transistor (MOSFET) in electronic circuits. Similar to a MOSFET, a nanofluidic transistor is the fundamental element for building a nanofluidic circuitry. There is possibility to achieve a nanofluidic circuitry, which is capable of logic operation and manipulation for ionic particles.

Since the conductance of ionic current flow is controlled by the gate voltage, using a material with high dielectric constant as the wall of the channel is desired. In this case, there is a stronger field seen within the channel due to a higher gate capacitance. A channel surface with a low surface charge is also desired in order to strengthen the effect of potential tuning by gate electrode. This increases the ability to spatially and temporally tune the ionic and electrostatic environment in the channel.

Field-Effect Reconfigurable Diode

By introducing an asymmetric field effect along the nanochannel, a field-effect reconfigurable nanofluidic diode is feasible, which features post-fabrication reconfiguration of the diode functions, such as the forward/reverse directions and the rectification degrees. Unlike the nanofluidic field-effect transistor, where only the amount of ions/molecules is regulated by an electrostatic potential, the field-effect reconfigurable diode can be used to control both directions and magnitudes of ion/molecule transport. This device could be deemed as the building blocks for ionic counterpart of the electronic field-programmable gate array.

Ionic Bipolar Transistors

Ionic bipolar transistors can be made from two conical channels with the smallest opening in nano-scaled dimension. By introducing opposite surface charges at each side, it is able to rectify ionic current as an ionic diode. An ionic bipolar transistor is built by combining two ionic diodes and forming a PNP junction along the inner surface of the channel. While the ionic current is from emitter end to collector end, the strength of the current can be modulated by the base electrode. The surface charge at the channel wall can be modified using chemical methods, by changing the electrolyte concentration or pH value.

Ionic Triodes

Nanofuidic triode is a three-terminal double junction nanofluidic device composed of positive-charged alumina and negative-charged silica nanochannels. The device is essentially a three-terminal bipolar junction transistor. By controlling the voltage across emitter and collector terminals, one can regulate the ion current from base terminal to one of the other two terminals, functioning as an ionic single-pole, double-throw switch.

Size Effect of Nanostructures

Width of Nanochannels

When surface charges present at the wall of a channel of micro-scaled width, counterions are attracted and co-ions are repelled by electrostatic force. The counterions form a shielding area near the wall. This region penetrate into solution to a certain distance called Debye length until the electric potential decays to the bulk value of neutrality. The Debye length is ranging typically from 1 nm to 100 nm for aqueous solutions.

In nano-channels, the Debye length is usually comparable with the channel width, therefore solution within the channel is charged. Ions inside the fluid is no longer shielded from surface charge. Instead, surface charge affect the dynamics of ions within a nano-channel.

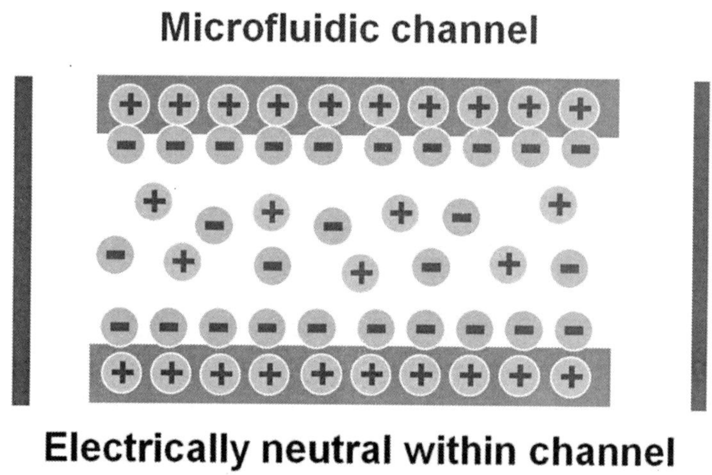

Figure: *Microfluidic channel, electrically neutral inside the channel*

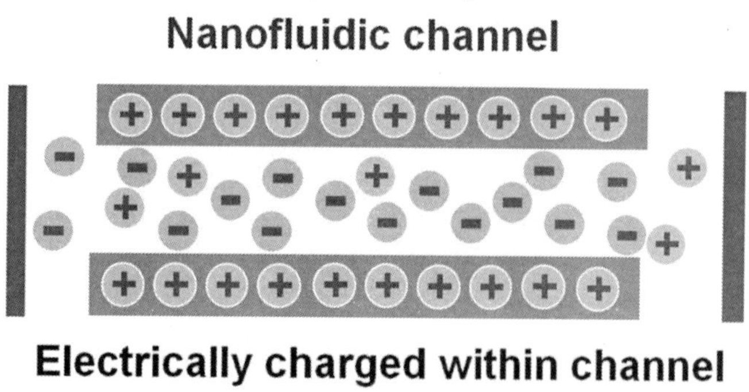

Figure: *Nanofluidic channel, electrically charged inside the channel*

Length of Nanochannels

It requires a channel to be narrow and long for it to have a good selectivity. In other words, a channel with a high aspect ratio has a better selectivity. To further increase its selectivity, it is required to have a highly charged wall.

The performance of ionic selectivity also largely related to the applied bias. With a low bias, a high selectivity is observed. With the increase of the bias voltage, there is an apparent decrease in the selectivity. For a nanochannel with a low aspect ratio, high selectivity is possible when the bias voltage is low.

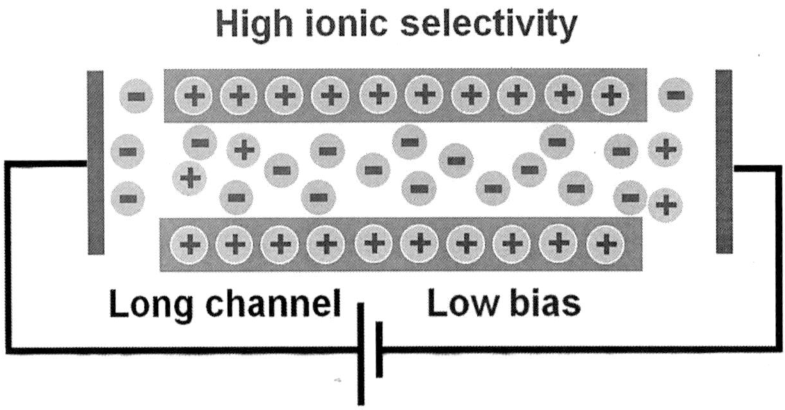

Figure: *Long channel and low bias voltage result in high selectivity*

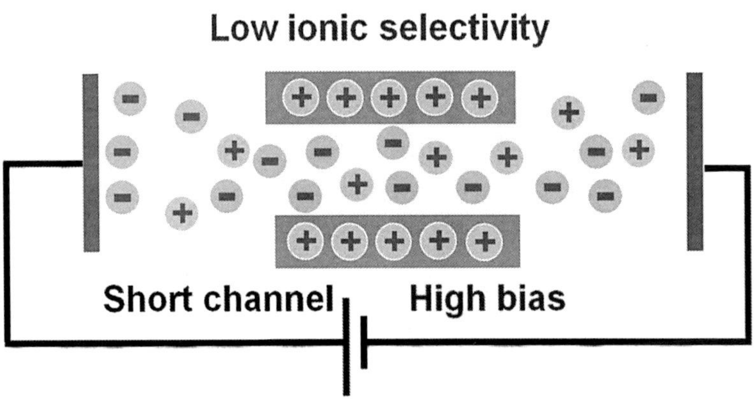

Figure: *Short channel and high bias voltage result in low selectivity*

Fabrication

The advantage of nanofluidic devices is from its feasibility to be integrated with electronic circuitry. Because they are built using the same manufacturing technology, it is possible to make a nanofluidic system with digital integrated circuit on a single chip. Therefore, the control and manipulation of particles in the electrolyte can be achieved in a real-time. Fabrication of nano-channels is categorized into top-down and bottom-up methods. Top-down methods are the conventional processes utilised in the IC industry and Microelectromechanical systems research. It begins with photolithograhpy on a bulk silicon wafer. Bottom-up methods, in contrast, starts with atoms or molecules with intrinsic nano-scaled dimension. By organise and combine these building blocks together, it is able to form a nanostructures as small as only a few nanometres.

Top-Down Methods

A typical method of top-down fabrication includes photolithography to define the geometry of channels on a substrate wafer. The geometry is created by several thin-film deposition and etching steps to form trenches. The substrate wafer is then bonded to another wafer to seal the trenches and form channels. Other technologies to fabricate nano-channels include surface micromachining with sacrificial layers, nano-imprinting lithography, and soft-lithography.

Bottom-up Methods

The most common method utilised for bottom-up fabrication is self-assembled monolayers (SAM). This method usually use biological materials to form a molecular monolayer on the substrate. Nano-channels can also be fabricated from the growth of carbon nanotubes (CNT) and quantum wires. The bottom-up methods usually give well-defined shapes with characteristic length about few nanometres. For these structures to be utilised as nanofluidic devices, the interconnection between nano-channels and microfluidic systems becomes an important issue.

There exist several ways to coat the inner surface with specific charges. Diffusion-limited patterning can be utilised because a bulk solution only penetrate the entrance of a nanochannel within a certain distance. Because the diffusion speed is different for each reactant. By introducing several steps of reactants flowing into the nanochannel, it is possible to pattern the surface with different surface charges inside the channel.

Application

Nanofluidic devices have been built for application in chemistry, molecular biology and medicine. The main purposes to use nanofluidic devices are separation and measurement of solutions containing nanoparticles for drug delivery, gene therapy and nanoparticle toxicology on a micro-total-analysis system. An important advantage of micro- and nano-scaled systems is the small amount of sample or reagent used in analysis. This reduces the time required for sample processing. It is also possible to achieve analysis in an array, which further speeds up processes and increases throughput of analysis.

Nanochannels are utilised to achieve single-molecule sensing and diagnosis, as well as DNA separation. In many cases, nanofluidic devices are integrated within a microfluidic system to facilitate logic operation of fluids. The future of nanofluidic systems will be focused on several areas such as analytical chemistry and biochemistry, liquid transport and metring, and energy conversion.

In nanofluidics, the valence numbers of the ions determines their net electrophoretic velocities. In other words, the velocity of an ion in the nano-channel is related not only to its ion mobility but also its ion valence. This enables the sorting function of nanofluidics, which cannot be done in a micro-channel. Therefore, it is possible to do sorting and separation for short strand DNA by using a nanochannel. For the single-molecule DNA application, the final goal is to sequence a strand of genomic DNA in a reproducible and precise result. Similar application can also be found in chromatography, or separation of various ingredients in the solution.

Application also can be found in synthesis of fibres. Polymer fibres can be created by electrospinning the monomers at an interface between liquid and vacuum. An organised polymer structure is formed from a flow of monomers aligning on a substrate.

There is also an attempt to bring nanofluidic technology into energy conversion. In this case, the electrical charged wall behaves as the stator, while the flowing solution as the rotor. It is observed that when the pressure-driven solvent flowing through a charged nanochannel, it can generate a streaming current and a streaming potential. This phenomenon can be used in electrical energy harvesting.

Advances in nanofabrication techniques and concerns about energy shortage make people interested in this idea. The main challenge is to increase efficiency, which is now only a few percent, compared with efficiencies of up to about 95 per cent for standard rotational electromagnetic generators.

Recent Advancements

Recent studies focus on the integration of nanofluidic devices into microsystems. An interface should be created for the connection between two length-scales. A system with solely nanofluidic devices standalone is impractical because it would requires a large driving pressure to make fluids flow into the nano-channel.

Nanofluidic devices are powerful in their high sensitivity and accurate manipulation of sample materials even down to a single molecule. Nevertheless, the drawback of nanofuidic separation systems is the relatively low sample throughput and its result in detection. One possible approach to deal with the problem is to use parallel separation channels with parallel detection in each channel. In addition, a better approach for detection needs to be created in view of the very small quantities of molecules present.

One of the biggest challenges in this research area are due to the peculiar size-effect. Researchers try to solve the problems caused by the extremely high surface-to-volume ratios. Under this condition, adsorption of molecules can lead to large losses and can also change the surface properties.

Another issue arises when the sample for detection is a relatively large molecule, such as DNA or protein. In the application for large molecule, clogging is a concern because the small size of the nanochannel makes it easy to happen. A low friction coating at inner surface of the channel is desired to avoid blocking of fluid channels in this application.

4

nanoHUB

nanoHUB.org is a science and engineering gateway comprising community-contributed resources and geared toward educational applications, professional networking, and interactive simulation tools for nanotechnology. Funded by the United States National Science Foundation (NSF), it is a product of the Network for Computational Nanotechnology (NCN), a multi-university initiative of eight member institutions including Purdue University, the University of California at Berkeley, the University of Illinois at Urbana-Champaign, Massachusetts Institute of Technology, the Molecular Foundry at Lawrence Berkeley National Laboratory, Norfolk State University, Northwestern University, and the University of Texas at El Paso. NCN was established to create a resource for nanoscience and nanotechnology via online services for research, education, and professional collaboration. NCN supports research efforts in nanoelectronics; nanoelectromechanical systems (NEMS); nanofluidics; nanomedicine, biology; and nanophotonics.

Infrastructure

The Network for Computational Nanotechnology was established in 2002. The US National Science Foundation (NSF) provided grants of about $14 million from 2002 through 2010, with principal investigator Mark S. Lundstrom.

The Web portal of NCN is nanoHUB.org and is an instance of a HUBZero hub. It offers simulation tools, course materials, lectures, seminars, tutorials, user groups, and online meetings. Interactive simulation tools are accessible from web browsers and run via a distributed computing network at Purdue University, as well as the TeraGrid and Open Science Grid. These resources are provided by hundreds of member contributors in the nanoscience community.

Resources include:

- Online Seminars
- Online Group Meeting Rooms
- Virtual Linux Workspaces that facilitate tool development within an in-browser Linux machine
- Online Workshops
- User Groups
- Interactive Simulation Tools for nanotechnology and related fields
- Lectures, Podcasts & Learning Materials in multiple formats
- Course Curricula for educators
- News & Events for Nanotechnology

Simulation Tools

The nanoHUB provides in-browser simulation tools geared toward nanotechnology, electrical engineering, chemistry, and semiconductor education. nanoHUB simulations are available to users as both stand-alone tools and part of structured teaching and learning curricula comprising numerous tools. Users develop and contribute their own tools for live deployment.

Examples of tools include:

SCHRED: calculates envelope wavefunctions and the corresponding bound-state energies in a typical Metal-Oxide-Semiconductor (MOS) or Semiconductor-Oxide-Semiconductor (SOS) structure and a typical SOI structure by solving self-consistently the one-dimensional (1D) Poisson equation and the 1D Schrödinger equation.

Quantum Dot Lab: computes the eigenstates of a particle in a box of various shapes including domes and pyramids.

Bulk Monte Carlo Tool: calculates the bulk values of the electron drift velocity, electron average energy and electron mobility for electric fields applied in arbitrary crystallographic direction in both column 4 (Si and Ge) and III-V (GaAs, SiC and GaN) materials.

Crystal Viewer: helps in visualizing various types of Bravais lattices, planes and Miller indices needed for many material, electronics and chemistry courses. Also large bulk systems for different materials (Silicon, InAs, GaAs, diamond, graphene, Buckyball) can be viewed using this tool.

Band Structure Lab: uses the sp3s*d5 tight binding method to compute E(k) for bulk, planar, and nanowire semiconductors. Using this tool, researchers may compute and visualize the band structures of bulk semiconductors, thin films, and nanowires for various materials, growth orientations, and strain conditions. Physical parameters such as the bandgap and effective mass can also be obtained from the computed E(k). The bandedges and effective masses of the bulk materials and the nanostructures structures can be analyzed as a function of various strain conditions.

Nano-Materials Simulation Toolkit: which uses molecular dynamics to simulate materials at the nano and micro-scale.

Ninithi: which can be used to visualize the 3D molecular geometries of graphene/nano-ribbons, carbon nanotubes and fullerenes. Ninithi also provides features to simulate the electronic band structures of graphene and carbon nanotubes.

Rappture Toolkit

The Rappture (Rapid APPlication infrastrucTURE) toolkit provides the basic infrastructure for the development of a large class of scientific applications, allowing scientists to focus on their core algorithm. It does so in a language-neutral fashion, so one may access Rappture in a variety of programming environments, including C/C++, Fortran and Python. To use Rappture, a developer describes all of the inputs and outputs for the simulator, and Rappture generates a Graphical User Interface (GUI) for the tool automatically.

Workspaces

A workspace is an in-browser Linux desktop that provides access to NCN's Rappture toolkit, along with computational resources available on the NCN, Open Science Grid, and TeraGrid networks. One can use these resources to conduct research, or as a development area for new simulation tools. One may upload code, compile it, test it, and debug it. Once code is tested and working properly in a workspace, it can be deployed as a live tool on nanoHUB.

A user can use normal Linux tools to transfer data into and out of a workspace. For example, sftp yourlogin@sftp.nanohub.org will establish a connection with a nanoHUB file share. Users can also use built-in WebDAV support on Windows, Macintosh, and Linux operating systems to access theier nanoHUB files on a local desktop.

Middleware

The web server uses a daemon to dynamically relay incoming VNC connections to the execution host on which an application session is running. Instead of using the port router to set up a separate channel by which a file import or export operation is conducted, it uses VNC to trigger an action on the browser which relays a file transfer through the main nanoHUB web server. The primary advantage of consolidating these capabilities into the web server is that it limits the entry point to the nanoHUB to one address. This simplifies the security model as well as reduces on the number of independent security certificates to manage.

One disadvantage of consolidating most communication through the web server is the lack of scalability when too much data is transferred by individual users. In order to avoid a network traffic jam, the web server can be replicated and clustered into one name by means of DNS round-robin selection.

The backend execution hosts that support Maxwell can operate with conventional Unix systems, Xen virtual machines, and a form of virtualization based on OpenVZ. For each system, a VNC server is pre-started for every session. When OpenVZ is used, that VNC server is started inside of a virtual container. Processes running in that container cannot see other processes on the physical system, dominate the resources of the physical machine, or make outbound network connections. By selectively overriding the restrictions imposed by OpenVZ, it is possible to synthesize a fully private environment for each application session that the user can use remotely.

Usage

The majority of users come from academic institutions using nanoHUB as part of their research and educational activities. Users also come from national labs and private industry. As a scientific resource, nanoHUB was cited hundrends of times in the scientific literature, peaking in 2009. Approximately sixty percent of the citations stem from authors not affiliated with the NCN. More than 200 of the citations refer to nanotechnology research, with more than 150 of them citing concrete resource usage. Twenty citations elaborate on nanoHUB use in education and more than 30 refer to nanoHUB as an example of national cyberinfrastructure.

Related Projects

The nanoHUB-U online course initiative was developed to enable students to study a subject in a 5-week framework roughly equivalent to a 1.0 credit class. No credit is given – quizzes and exams are simple and are intended to be aids to learning rather than rigorous tests for acquired skills. In the spirit of a research university, nanoHUB-U courses aim to bring new advances and understanding from research into the curriculum. Every effort is made to present courses in a way that is accessible to beginning graduate students with a variety of different backgrounds. What this means in practice is that the number of prerequisites should be kept to an absolute minimum. The ideal nanoHUB-U course is accessible to any students with an undergraduate degree in engineering or the physical sciences.

Nanometrology

Nanometrology is a subfield of metrology, concerned with the science of measurement at the nanoscale level. Nanometrology has a crucial role in order to produce nanomaterials and devices with a high degree of accuracy and reliability in nanomanufacturing.

A challenge in this field is to develop or create new measurement techniques and standards to meet the needs of next-generation advanced manufacturing, which will rely on nanometre scale materials and technologies. The needs for measurement and characterization of new sample structures and characteristics far exceed the capabilities of current measurement science. Anticipated advances in emerging U.S. nanotechnology industries will require revolutionary metrology with higher resolution and accuracy than has previously been envisioned.

Control of the critical dimensions are the most important factors in nanotechnology. Nanometrology today, is to a large extent based on the development in semiconductor technology. Nanometrology is the science of measurement at the nanoscale level. Nanometre or nm is equivalent to 10^-9 m. In Nanotechnology accurate control of dimensions of objects is important. Typical dimensions of nanosystems vary from 10 nm to a few hundred nm and while fabricating such systems measurement up to 0.1 nm is required.

At nanoscale due to the small dimensions various new physical phenomena can be observed. For example, when the crystal size is smaller than the electron mean free path the conductivity of the crystal changes. Another example is the discretization of stresses in the system. It becomes important to measure the physical parameters so as to apply these phenomena into engineering of nanosystems and manufacturing them. The measurement of length or size, force, mass, electrical and other properties is included in Nanometrology. The problem is how to measure these with reliability and accuracy. The measurement techniques used for macro systems cannot be directly used for measurement of parameters in nanosystems. Various techniques based on physical phenomena have been developed which can be used for measure or determine the parameters for nanostructures and nanomaterials. Some of the popular ones are X-Ray Diffraction, Transmission Electron Microscopy, High Resolution Transmission Electron Microscopy, Atomic Force Microscopy, Scanning Electron Microscopy, Field Emission Scanning Electron Microscopy and Brunauer, Emmett, Teller method to determine specific surface.

Nanotechnology is an important field because of the large number of applications it has and it has become necessary to develop more precise techniques of measurement and globally accepted standards. Hence progress is required in the field of Nanometrology.

Development Needs

Nanotechnology can be divided into two branches. The first being Molecular Nanotechnology which involves bottom up manufacturing and the second is Engineering Nanotechnology which involve the development and processing of materials and systems at nanoscale. The measurement and manufacturing tools and techniques required for the two branches are slightly different.

Furthermore, Nanometrology requirements are different for the industry and research institutions. Nanometrology of research has progressed faster than that for industry mainly because implementing nanometrology for industry is difficult. In research oriented nanometrology resolution is important whereas in industrial nanometrology accuracy is given precedence over resolution. Further due to economic reasons it is important to have low time costs in industrial nanometrology it is not important for research nanometrology. The various measurement techniques available today require a controlled environment like in vacuum, vibration and noise free environment. Also, in industrial nanometrology requires that the measurements be more quantitative with minimum number of parameters.

Standards

International Standards: Metrology standards are objects or ideas that are designated as being authoritative for some accepted reason. Whatever value they possess is useful for comparison to unknowns for the purpose of establishing or confirming an assigned value based on the standard. The execution of measurement comparisons for the purpose of establishing the relationship between a standard and some other measuring device is calibration. The ideal standard is independently reproducible without uncertainty. The world-wide market for products with nanotechnology applications is projected to be at least a couple of hundred billion dollars in the near future. Until recently, there almost no established internationally accepted standards for nanotechnology related field.

The International Organisation for Standardization TC-229 Technical Committee on Nanotechnology recently published few standards for terminology, characterization of nanomaterials and nanoparticles using measurement tools like AFM, SEM, Interferometres, optoacoustic tools, gas adsorption methods etc.

Certain standards for standardization of measurements for electrical properties have been published by the International Electrotechnical Commission. Some important standards which are yet to be established are standards for measuring thickness of thin films or layers, characterization of surface features, standards for force measurement at nanoscale, standards for characterization of critical dimensions of nanoparticles and nanostructures and also Standards for measurement for physical properties like conductivity, elasticity etc.

National Standards

Because of the importance of nanotechnology in the future, countries around the world have programmes to establish national standards for nanometrology and nanotechnology. These programmes are run by the national standard agencies of the respective countries. In the United States, National Institute of Standards and Technology has been working on developing new techniques for measurement at nanoscale and has also established some national standards for nanotechnology. These standards are for nanoparticle characterization, Roughness Characterization, magnification standard, calibration standards and etc.

Calibration

It is difficult to provide samples using which precision instruments can be calibrated at nanoscale. Reference or Calibration standards are important for repeatability to be ensured. But there are no international standards for calibration and the calibration artefacts provided by the company along with their equipment is only good for calibrating that particular equipment. Hence it is difficult to select a universal calibration artefact using which we can achieve repeatability at nanoscale. At nanoscale while calibrating care needs to be taken for influence of external factors like vibration, noise, motion and internal factors like the interaction between the artefact and the equipment which can cause significant deviations.

Measurement Techniques

In the last 70 years various techniques for measuring at nanoscale have been developed. Most of them based on some physical phenomena observed on particle interactions or forces at nanoscale. Some of the most commonly used techniques are Atomic Force Microscopy, X-Ray Diffraction, Scanning Electron Microscopy, Transmission Electron Microscopy, High Resolution Transmission Electron Microscopy, and Field Emission Scanning Electron Microscopy.

Atomic Force Microscopy (AFM) is one of the most common measurement techniques. It can be used to measure Topology, grain size, frictional characteristics and different forces. It consists of a silicon cantilever with a sharp tip with a radius of curvature of a few nanometres. The tip is used as a probe on the specimen to be measured. The forces acting at the atomic level between the tip and the surface of the specimen cause the tip to deflect and this deflection is detected using a laser spot which is reflected to an array of photodiodes.

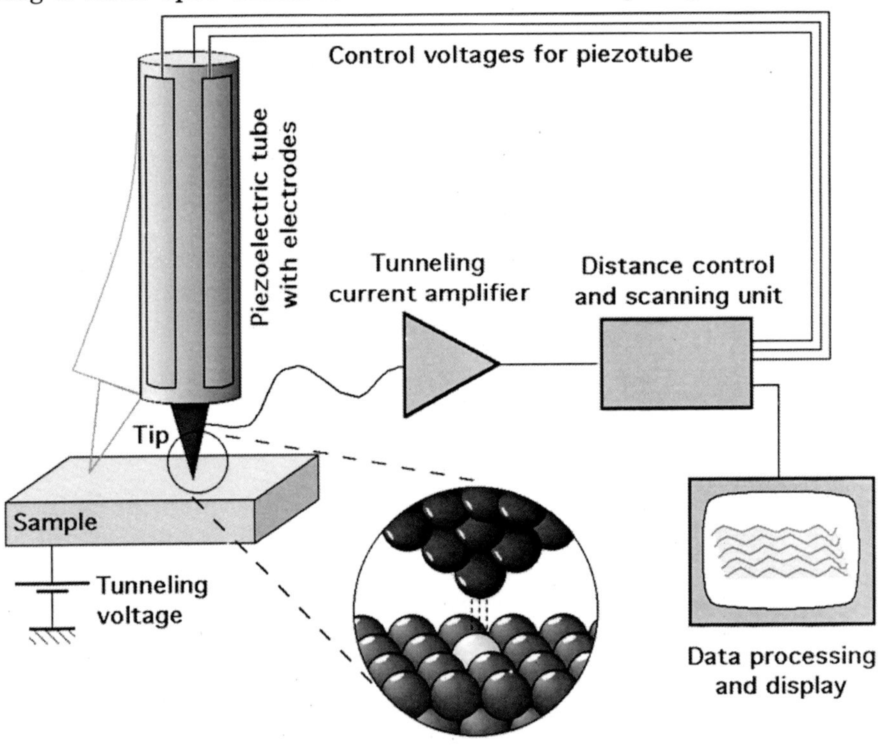

***Figure:** Diagram of* Scanning Tunneling Microscope.

Scanning Tunneling Microscopy (STM) is another instrument commonly used. It is used to measure 3-D topology of the specimen. The STM is based on the concept of quantum tunneling. When a conducting tip is brought very near to the surface to be examined, a bias (voltage difference) applied between the two can allow electrons to tunnel through the vacuum between them. Measurements are made by monitoring the current as the tip's position scans across the surface, which can then be used to display an image.

Another commonly used instrument is the Scanning Electron Microscopy (SEM) which apart from measuring the shape and size of the particles and topography of the surface can be used to determine the composition of elements and compounds the sample is composed of. In SEM the specimen surface is scanned with a high energy electron beam. The electrons in the beam interact with atoms in the specimen and interactions are detected using detectors. The interactions produced are back scattering of electrons, transmission of electrons, secondary electrons etc. To remove high angle electrons magnetics lenses are used.

The instruments mentioned above produce realistic pictures of the surface are excellent measuring tools for research. Industrial applications of nanotechnology require the measurements to be produced need to be more quantitative. The requirement in industrial nanometrology is for higher accuracy than resolution as compared to research nanometrology.

Nano Coordinate Measuring Machine

A Coordinate Measuring Machine (CMM) that works at the nanoscale would have a smaller frame than the CMM used for macroscale objects. This is so because it may provide the necessary stiffness and stability to achieve nanoscale uncertainties in x,y and z directions. The probes for such a machine need to be small to enable a 3-D measurement of nanometre features from the sides and from inside like nanoholes. Also for accuracy laser interferometres need to be used. NIST has developed a surface measuring instrument, called the Molecular Measuring Machine. This instrument is basically an STM. The x- and y-axes are read out by laser interferometres. The molecules on the surface area can be identified individually and at the same time the distance between any two molecules can be determined. For measuring with molecular resolution, the measuring times become very large for even a very small surface area. Ilmenau Machine is another nanomeasuring machine developed by researchers at the Ilmenau University of Technology.

Components of a Nano CMM

- Nanoprobes
- Control Hardware

- 3D-nanopositioning platform
- Instruments with high resolution and accuracy for linear measurement
- Instruments with high resolution and accuracy for angular measurement

Traceability

In metrology at macro scale achieving traceability is quite easy and artefacts like scales, laser interferometres, step gauges, and straight edges are used. At nanoscale a crystalline mica or silicon surface is considered suitable used as calibration artefact for achieving traceability. But it is not always possible to ensure traceability. Like what is a straight edge at nanoscale and even if take the same standard as that for macroscale there is no way to calibrate it accurately at nanoscale. This so because the requisite internationally or nationally accepted reference standards are not always there. Also the measurement equipment required to ensure traceability has not been developed. The generally used for traceability are miniaturisation of traditional metrology standards hence there is a need for establishing nanoscale standards. Also there is a need to establish some kind of uncertainty estimation model. Traceability is one of the fundamental requirements for manufacturing and assembly of products when multiple producers are there.

Tolerance

Tolerance is the permissible limit or limits of variation in dimensions, properties, or conditions without significantly affecting functioning of equipment or a process. Tolerances are specified to allow reasonable leeway for imperfections and inherent variability without compromising performance. In nanotechnology the systems have dimensions in the range of nanometres. Defining tolerances at nanoscale with suitable calibration standards for traceability is difficult for different nanomanufacturing methods. There are various integration techniques developed in the semiconductor industry that are used in nanomanufacturing.

Integration Techniques

- In Hetero integration direct fabrication of nanosystems from compound substrates is done. Geometric tolerances are required to achieve the functionality of the assembly.
- In Hybrid integration nanocomponents are placed or assembled on a substrate fabricating functioning nanosystems. In this

technique, the most important control parameter is the positional accuracy of the components on the substrate.

- In Monolithic integration all the fabrication process steps are integrated on a single substrate and hence no mating of components or assembly is required. The advantage of this technique is that the geometric measurements are no longer of primary importance for achieving functionality of nanosystem or control of the fabrication process.

Classification of Nanostructures

There are a variety of Nanostructures like nanocomposites, nanowires, nanopowders, nanotubes, fullerenes nanofibres, nanocages, nanocrystallites, nanoneedles, nanofoams, nanomeshes, nanoparticles, nanopillars, thin films, nanorods, nanofabrics, quantumdots etc. The most common way to classify nano structures is by their dimensions.

Dimensional Classification

Dimensions	*Criteria*	*Examples*
Zero-dimensional (0-D)	The nanostructure has all dimensions in the nanometer range.	Nanoparticles, quantum dots, nanodots
One-dimensional (1-D)	One dimension of the nanostructure is outside the nanometer range.	Nanowires, nanorods, nanotubes
Two-dimensional (2-D)	Two dimensions of the nanostructure are outside the nanometer range.	Coatings, thin-film-multilayers
Three-dimensional (3-D)	Three dimensions of the nanostructure are outside the nanometer range.	Bulk

Classification of Grain Structure

Nanostructures can be classified on the basis of the grain structure and size there are made up of. This is applicable in the cas of 2-dimensional and 3-Dimensional Nanostructurs.

Surface Area Measurement

For nanopowder to determine the specific surface area the B.E.T. method is commonly used. The drop of pressure of nitrogen in a closed container due to adsorption of the nitrogen molecules to the surface of the material inserted in the container is measured. Also, the shape of the nanopowder particles is assumed to be spherical.

$$D = 6/(r*A)$$

Where "D" is the effective diameter, "ρ" is the density and "A" is the surface area fond from the B.E.T. method.

Nanonetwork

A nanonetwork or nanoscale network is a set of interconnected nanomachines, (devices a few hundred nanometres or a few micrometres at most in size), which are able to perform only very simple tasks such as computing, data storing, sensing and actuation. Nanonetworks are expected to expand the capabilities of single nanomachines both in terms of complexity and range of operation by allowing them to coordinate, share and fuse information. Nanonetworks enable new applications of nanotechnology in the biomedical field, environmental research, military technology and industrial and consumer goods applications.

Communication Approaches

Classical communication paradigms need to be revised for the nanoscale. The two main alternatives for communication in the nanoscale are based either on electromagnetic communication or on molecular communication.

Electromagnetic

This is defined as the transmission and reception of electromagnetic radiation from components based on novel nanomaterials. Recent advancements in carbon and molecular electronics have opened the door to a new generation of electronic nanoscale components such as nanobatteries, nanoscale energy harvesting systems, nano-memories, logical circuitry in the nanoscale and even nano-antennas. From a communication perspective, the unique properties observed in nanomaterials will decide on the specific bandwidths for emission of electromagnetic radiation, the time lag of the emission, or the magnitude of the emitted power for a given input energy, amongst others.

For the time being, two main alternatives for electromagnetic communication in the nanoscale have been envisioned. First, it has been experimentally demonstrated that is possible to receive and demodulate an electromagnetic wave by means of a nanoradio, i.e., an electromechanically resonating carbon nanotube which is able to decode an amplitude or frequency modulated wave. Second, graphene-based nano-antennas have been analyzed as potential electromagnetic radiators in the Terahertz band.

Molecular

Molecular communication is defined as the transmission and reception of information by means of molecules. The different molecular communication techniques can be classified according to the type of molecule propagation in walkaway-based, flow-based or diffusion-based communication. In walkway-based molecular communication, the molecules propagate through pre-defined pathways by using carrier substances, such as molecular motors. This type of molecular communication can also be achieved by using E. coli bacteria as chemotaxis.

In flow-based molecular communication, the molecules propagate through diffusion in a fluidic medium whose flow and turbulence are guided and predictable. The hormonal communication through blood streams inside the human body is an example of this type of propagation. The flow-based propagation can also be realised by using carrier entities whose motion can be constrained on the average along specific paths, despite showing a random component. A good example of this case is given by pheromonal long range molecular communications.

In diffusion-based molecular communication, the molecules propagate through spontaneous diffusion in a fluidic medium. In this case, the molecules can be subject solely to the laws of diffusion or can also be affected by non-predictable turbulence present in the fluidic medium. Pheromonal communication, when pheromones are released into a fluidic medium, such as air or water, is an example of diffusion-based architecture. Other examples of this kind of transport include calcium signalling among cells, as well as quorum sensing among bacteria.

Based on the macroscopic theory of ideal (free) diffusion the impulse response of a unicast molecular communication channel was reported in a paper that identified that the impulse response of the ideal diffusion based molecular communication channel experiences temporal spreading. Such temporal spreading has a deep impact in the performance of the system e.g. in creating the intersymbol interference (ISI) at the receiving nanomachine.

In order to detect the concentration-encoded molecular signal two detection methods named sampling-based detection (SD) and energy-based detection (ED) have been proposed. While the SD approach is based on the concentration amplitude of only one sample taken at a suitable time instant during the symbol duration, the ED approach is based on the total accumulated number of molecules received during the entire symbol duration. In order to reduce the impact of ISI a controlled pulse-width based molecular communication scheme has been analysed. The work presented in showed that it is possible to realise multilevel amplitude modulation based on ideal diffusion. A comprehensive study of pulse-based binary and sinus-based, concentration-encoded molecular communication system have also been investigated.

IEEE P1906.1

The P1906.1 - Recommended Practice for Nanoscale and Molecular Communication Framework is a standards working group in the sponsored by IEEE Communications Society Standards Development Board whose goal is to develop a common framework for nanoscale and molecular communications. Because this is an emerging technology, the framework is meant to encourage innovation by determining a common definition, goals, use-cases, and framework allowing the technology to advance at a faster rate.

Membership

Membership includes industry and academia with backgrounds in mathematical modelling, engineering, physics, economics and biological sciences.

Definition

The precise definition of a nanoscale communication network is under development with respect to three abstract aspects:

1. The portions of the system that must exist at the nanoscale
2. The components of the signal and its transmission
3. Speed of communication

Electronic components such as transistors, or electrical/electromagnetic message carriers whose operation is similar at the macroscale and nanoscale are excluded from the definition. A human-engineered, synthetic component must form part of the system because it is important to avoid standardizing nature or physical processes.

The definition of communication, particularly in the area of cell-surface interactions as viewed by biologists versus non-biologists has been a topic of debate. The interface is viewed as a communication channel, whereas the 'receptors-signalling-gene expression' events are the network.

Applications

Applications are numerous, however, there appears to be strong emphasis on medical and biological use-cases in nanomedicine.

Nanotechnology in Water Treatment

Nanotechnology, the engineering and art of manipulating matter at the nanoscale (1-100 nm), offers the potential of novel nanomaterials for the treatment of surface water, groundwater and wastewater contaminated by toxic metal ions, organic and inorganic solutes and microorganisms. Due to their unique activity toward recalcitrant contaminants many nanomaterials are under active research and development for use in the treatment of water.

The present market of nanotech-based technologies applied in water treatment consists of reverse osmosis, nanofiltration, ultrafiltration membranes. Indeed, among emerging products one can name nanofibre filters, carbon nanotubes and various nanoparticles. Nanotechnology is expected to deal more efficiently with contaminants which convectional water treatment systems struggle to treat, including bacteria, viruses and heavy metals. This efficiency generally stems from the very high specific surface area of nanomaterials which increases dissolution, reactivity and sorption of contaminants.

Detection of Microbial Pathogens

To maintain public health, pathogens in water need to be identified rapidly and reliably. Unfortunately, traditional laboratory culture tests take days to complete. Faster methods involving enzymes, immunological or genetic tests are under development.

Nanofibres and Nanobiocides

Water filtration may be improved with the use of nanofibre membranes and the use of nanobiocides, which appear promisingly effective.

Biofilm Removal

Biofilms are mats of bacteria wrapped in natural polymers. These can be difficult to treat with antimicrobials or other chemicals. They can be cleaned up mechanically, but at the cost of substantial down-time and labour. Work is in progress to develop enzyme treatments that may be able to break down such biofilms.

Groundwater Remediation Applications of Nanotechnology

In groundwater remediation a variety of technologies may be used for decontamination. Nanotechnology, especially using zero-valent metals (ZVMs), represents just one of these approaches.

Remediation Process

Nanotechnology offers the ability to effectively enable contaminant treatment *in situ*. The process begins with the injection of nanoparticles into a contaminated aquifer via an injection well. The nanoparticles are then transported to the source of contamination by the groundwater flow where they then degrade the contaminant. Nanoparticles can sequester (via adsorption or complexation), immobilizing them, or they can degrade the contaminants to less harmful compounds. Contaminant transformations are typically redox reactions. When the nanoparticle is the oxidant or reductant, it is considered reactive.

Reaction Products

When using *in situ* remediation the reactive products must be considered for two reasons. One reason is that a reactive product might be more harmful or mobile than the parent compound. Another reason is that the products can affect the effectiveness and/or cost of remediation. TCE (trichloroethylene), under reducing conditions by nanoiron, may sequentially dechlorinate to DCE (dichloroethene) and VC (vinyl chloride). VC is known to be more harmful than TCE, meaning this process would be undesirable.

Delivery and Transport Issues

Using nanomaterials for groundwater remediation is promising due to the availability and effectiveness of many nanomaterials for degrading or sequestering contaminants. Their ability to be used *in situ* also makes them very attractive: there is no need for excavation or pumping them out of the ground. In order to realise the potential of nanomaterials the ability to inject them to the subsurface and transport them to the contaminant source is imperative. Unless they are delivered to and stay at the contaminant source, they cannot be used efficiently.

Injection Methods

Reactive nanoparticles can be injected into a well where they will then be allowed to transport down gradient to the contaminated area. Drilling and packing a well is quite expensive. Direct push wells are lower cost than drilled wells and are the most often used delivery tool for remediation with nanoiron. A nanoparticle slurry can be injected along the vertical range of the probe to provide treatment to specific aquifer regions.

Reasons for Success

There are two main reasons for the capabilities of nanoparticles as a remediation method. First is the small size, which grants them access to contaminants sorbed to soil. Also, their movement is largely governed by Brownian motion as compared to gravity. Thus, the flow of groundwater is sufficient to transport the particles. Nanoparticles then can remain in suspension to establish an *in situ* treatment zone.

Preparation

Nanoparticles can be prepared by using sodium borohydride as the key reductant. $NaBH_4$ (0.2 M) is added into $FeCl_3 \cdot 6H_2$ (0.05 M) solution (~1:1 volume ratio). Ferric iron is reduced via the following reaction:

$4Fe^{3+}$ + 3BH–

4 + $9H_2O \rightarrow 4Fe^0 + 3H_2BO$–

3 + $12H^+ + 6H_2$

Palladized Fe particles are prepared by soaking the nanoscale iron particles with an ethanol solution of 1wt% of palladium acetate ($[Pd(C_2H_3O_2)2]_3$). This causes the reduction and deposition of Pd on the Fe surface:

$$Pd^{2+} + Fe^0 \rightarrow Pd^0 + Fe^{2+}$$

Similar methods may be used to prepared Fe/Pt, Fe/Ag, Fe/Ni, Fe/ Co, and Fe/Cu bimetallic particles. With the above methods, nanoparticles of diameter 50-70 nm may be produced. The average specific surface area of Pd/Fe particles is about 35 m^2/g. Ferrous iron salt has also been successfully used as the precursor.

Titanium Oxide

TiO_2 has been widely used for water treatment as it has a wide band gap energy (3.2 eV), it is inexpensive, non-toxic, highly photoactive and insoluble in water. A study of the degradation of Reactive Brilliant X-3B (a reactive dye) and catechol (a refractory organic compound) has been performed. The degradation occurred via oxidation with ozone under UV light under these conditions: a) without a catalyst, b) with TiO_2 as a catalyst, and c) with carbon black-coated TiO_2 catalyst on Al thin films. The reaction rate almost doubled with the use of TiO_2 nanoparticles and attacking carbon black TiO_2 made the separation from the pollutant easy and affordable.

In the degradation of sucrose and nitrate over titania-coated nano-hematite catalysts, the presence of Fe^{3+} in hematite results in trapping the electrons more efficiently than with TiO_2 alone, reducing the degradation time from more than 60 min to about 45 min. TiO_2 coupled with different metal oxides kept the e^- trapped longer, reducing the recombination rate of the electrons to their hydrogen ions, and improved photocatalytic activity in degradation of methyl orange as an organic pollutant.

Photodecontamination

Decontamination of wastewater is one of the most successful photochemical applications of photons. Solar photocatalysis using TiO_2 nanocrystals removed various refractory pollutants, such as phonls, 4-chlorophenol, nitrobenzene, 2-chlorobenzoic acid and hydrobutanedioic acid. Solar light with TiO_2 had maximum removal efficiency in comparison to both UV and solar radiation. However, some study have shown that UV radiation in more effective in removing coloured dye.

Solid-Phase Microextraction

In environmental problems with trace contaminants, such as groundwater contamination, there is often the need for rapid, portable, cost-effective measurement systems for these trace contaminants. Unfortunately, instruments that can operate outside of a laboratory often are not sensitive enough for detection. One way around this is to separate the analyte from the sample and concentrate them to a smaller volume, easing detection and measurement. When this is done with small quantities of solid sorbents it is referred to as solid-phase microextraction (SPME). Typically, effective SPME materials require high affinities for the target analyte and large surface areas.

Self-Assembled Monolayers on Mesoporous Supports

Self-assembled monolayers on mesoporous supports (SAMMS) have numerous properties that make them ideal sorbents for SPME. The mesoporous silica structure is made through a surfactant templated sol-gel process that endows SAMMS with high surface areas and a rigid open pore structure. SAMMS have been shown to be excellent sorbents for many targets, including heavy metals such as mercury, lead, and cadmium, chromate and arsenate, and radionuclides such as ^{99}Tc, ^{137}CS, uranium, and the actinides.

Nano-Thermite

A Nano-thermite or "super-thermite" is a metastable intermolecular composite (MICs) characterized by a particle size of its main constituents, a metal and a metal oxide, under 100 Nanometres. This allows for high and customizable reaction rates. Nano-thermites contain an oxidizer and a reducing agent, which are intimately mixed on the nanometre scale. MICs, including nano-thermitic materials, are a type of reactive materials investigated for military use, as well as for general applications involving propellants, explosives, and pyrotechnics.

What distinguishes MICs from traditional thermites is that the oxidizer and a reducing agent, normally iron oxide and aluminium, are in the form of extremely fine powders (nanoparticles). This dramatically increases the reactivity relative to micrometre-sized powder thermite. As the mass transport mechanisms that slow down the burning rates of traditional thermites are not so important at these scales, the reactions become kinetically controlled and proceed much more quickly.

Potential Uses

Historically, pyrotechnic or explosive applications for traditional thermites have been limited due to their relatively slow energy release rates. Because nanothermites are created from reactant particles with proximities approaching the atomic scale, energy release rates are far greater.

MICs or Super-thermites are generally developed for military use, propellants, explosives, and pyrotechnics. Research into military applications of nano-sized materials began in the early 1990s.

Because of their highly increased reaction rate, nanosized thermitic materials are being studied by the U.S. military with the aim of developing new types of bombs several times more powerful than conventional explosives. Nanoenergetic materials can store more energy than conventional energetic materials and can be used in innovative ways to tailor the release of this energy. Thermobaric weapons are one potential application of nanoenergetic materials.

Types

There are many possible thermodynamically stable fuel-oxidizer combinations. Some of them are:

- Aluminium-molybdenum(VI) oxide
- Aluminium-copper(II) oxide
- Aluminium-iron(II,III) oxide
- Antimony-potassium permanganate
- Aluminium-potassium permanganate
- Aluminium-bismuth(III) oxide
- Aluminium-tungsten(VI) oxide hydrate
- Aluminium-fluoropolymer (typically Viton)
- Titanium-boron (burns to titanium diboride)

In military research, aluminium-molybdenum oxide, aluminium-Teflon and aluminium-copper(II) oxide have received considerable attention. Other compositions tested were based on nanosized RDX and with thermoplastic elastomers. PTFE or other fluoropolymer can be used as a binder for the composition. Its reaction with the aluminium, similar to magnesium/teflon/viton thermite, adds energy to the reaction. Of the listed compositions, that with potassium permanganate has the highest pressurization rate.

Nanoparticles can be prepared by spray drying from a solution, or in case of insoluble oxides, spray pyrolysis of solutions of suitable precursors. The composite materials can be prepared by sol-gel techniques or by conventional wet-mixing and pressing.

Similar but not identical are nano-laminated pyrotechnic compositions, or energetic nanocomposites, in which fuel and oxidizer are deposited alternately in thin layers. The materials and the thickness of the layers control the detonating properties.

Production

A method for producing nanoscale, or ultra fine grain (UFG) aluminium powders, a key component of most nano-thermitic materials, is the dynamic gas-phase condensation method, pioneered by Wayne Danen and Steve Son at Los Alamos National Laboratory. A variant of the method is being used at the Indian Head Division of the Naval Surface Warfare Centre. Another production method for nanoaluminium powder is the pulsed plasma process developed by NovaCentrix (formerly Nanotechnologies). The powders made by both processes are indistinguishable. A critical aspect of the production is the ability to produce particles of sizes in the tens of nanometre range, as well as with a limited distribution of particle sizes. In 2002, the production of nano-sized aluminium particles required considerable effort, and commercial sources for the material were limited. Current production levels are now beyond 100 kg/month.

An application of the sol-gel method, developed by Randall Simpson, Alexander Gash and others at the Lawrence Livermore National Laboratory, can be used to make the actual mixtures of nanostructured composite energetic materials. Depending on the process, MICs of different density can be produced. Highly porous and uniform products can be achieved by supercritical extraction.

Ignition

Nanoscale composites are easier to ignite than traditional thermites. A nichrome bridgewire can be used in some cases. Other means of ignition can include flame or laser pulse. Los Alamos National Laboratory (LANL) is developing super-thermite electric matches that use comparatively low ignition currents and resist friction, impact, heat and static discharge. MICs have been investigated as a possible replacement for lead (e.g. lead styphnate, lead azide) in percussion caps and electric matches. Compositions based on Al-Bi_2O_3 tend to be used. PETN may be optionally added. MICs can be also added to high explosives to modify their properties. Aluminium is typically added to explosives to increase their energy yield. Addition of small amount of MIC to aluminium powder increases overall combustion rate, acting as a burn rate modifier.

The products of a thermite reaction, resulting from ignition of the thermitic mixture, are usually metal oxides and elemental metals. At the temperatures prevailing during the reaction, the products can be solid, liquid or gaseous, depending on the components of the mixture. Super-thermite electric matches developed by LANL can create simple sparks, hot slag, droplet, or flames as thermal-initiating outputs to ignite other incendiaries or explosives.

Hazards

Like conventional thermite, super thermite reacts at very high temperature and is difficult to extinguish. The reaction produces dangerous ultra-violet (UV) light requiring that the reaction not be viewed directly, or that special eye protection (for example, a welder's mask) be worn. The reactivity of a nanolaminate can vary, possibly making it more sensitive than thermite. In addition, super thermites are very sensitive to electrostatic discharge (ESD). Surrounding the metal oxide particles with carbon nanofibres may make nanothermites safer to handle.

Wet Nanotechnology

Wet nanotechnology (also known as wet nanotech) involves working up to large masses from small ones. Wet nanotechnology requires water in which the process occurs. The process also involves chemists and biologists trying to reach larger scales by putting together individual molecules. While Eric Drexler put forth the idea of nano-assemblers working dry, wet nanotech appears to be the likely first area in which something like a nano-assembler may achieve economic results. Pharmaceuticals and bioscience are central features of most nanotech start-ups. Richard A.L. Jones calls nanotechnology that steals bits of natural nanotechnology and puts them in a synthetic structure *biokleptic nanotechnology*. He calls building with synthetic materials according to nature's design principles *biomimetic nanotechnology*. Using these guiding principles could lead to trillions of nanotech robots, that resemble bacteria in structural properties, entering a person's blood stream to do medical treatments.

Background

Wet nanotechnology is a fairly new area of nanotech that is going to mostly be dominated by the different forms of wet engineering. The processes that will be used are going to take place in aqueous solutions and are very close to that of bio-molecular manufacturing; this deals greatly with the creation of things like proteins found in our body such as DNA. In order for mankind to mimic these nanoscale machines in a way that they could be produced with some efficiency, they must look into bottom-up manufacturing. Bottom-up manufacturing deals with manipulating individual atoms during the manufacturing process, so that there is absolute control of their placement and interactions. Then from the atomic scale, nanomachines could be made and even be designed to self-replicate themselves as long as they are designed in an environment with copious amount of the needed materials. Because individual atoms are being manipulated in the process, bottom-up manufacturing is often referred to as "atom by atom" manufacturing. If the manufacturing of nanomachines can be made more readily available through improved techniques, there could be a large economic and social impact. This would start with improvements in making microelectromechanical systems and then would allow for the creation of nanoscale biological sensors along with things that have not been thought of yet. This is because "wet" nanotech is only in the beginning if its life. Scientists and engineers alike feel that biomimetics is a great way to start looking at creating nanoscale machines. Humans have only had a few thousand years to try to learn about the mechanics of things at really small scales. However, nature has been working on perfecting the design and functionality of nanomachines for millions of years. This is why there are already nanomachines, such as ATP synthase, working in our bodies that have on unheard of 95% efficiency.

"Wet" vs. "Dry"

Wet nanotechnology is a form of wet engineering as opposed to dry engineering. There are different fields that deal with those two types of engineering. Molecular scientists, genetic engineers, and biologists all deal with wet engineering. They study processes that happen in life, and for the most part those processes take place in aqueous environments. Our bodies are made up mostly of water. Electrical and mechanical engineers are on the other side of the line in dry engineering. They are involved with processes and manufacturing that does not occur in aqueous environments. For the most part, wet engineering deals with "soft" materials that allow for flexibility which is vital at the nanoscale in biological manufacturing. Dry engineers mostly handle things with rigid structures and parts. These differences stem for the fact that the forces that the two types of engineers must deal with are very different. At a larger scale, most things are dominated by Newtonian physics. However, when one looks at the nanoscale, especially in biological matters, the dominating force is Brownian motion. Because nanotechnology in the new age is going to most likely deal with both dry and wet in conjunction with each other, there is going to have to be a change in the way society looks at engineering and manufacturing. People will have to be not only well educated in engineering but also in biology because the integration of the two is how there will be the largest improvements in nanotechnology.

Brownian Motion as It Relates to Wet Nanotech

With the existence of natural nanomachines, "a complex precision microscopic-sized machine that fits the standard definition of a machine", such as ATP synthase and T4 bacteriophage, scientists and biologists know that they are capable of making similar types of machines at the same scale. However, nature has had a long time to perfect the building and creation of these nanomachines and humankind has only just begun to look into them with greater interest. This interest may have been sparked because of the existence of nanomachines such as ATP synthase (adenosine triphosphate), which is the "second in importance only to DNA". ATP is the main energy converter that our bodies contain and without it, life as we know it would not be able to flourish or even survive.

What does Brownian Motion have to do with Complex Nanomachines?

Brownian motion is a random, constantly fluctuating force that acts on a body in environments that are at a microscale. This force is one that mechanical engineers and physicists are not used to dealing with because, at the larger scale that humankind tends to think of things, this force is not one that needs to be taken into account. People think of gravity, inertia, and other physics based forces that act on us all the time, however at the nanoscale those forces are mostly "negligible". In order for nanomachines to be recreated by man, either there will need to be discoveries that allow us to understand how to "exploit" Brownian motion as nature does or find a way to work around it by using materials that are rigid enough to stand up to these forces. The way that nature has been able to exploit Brownian motion is through self-assembly. This force pushes and pulls all of the proteins and amino acids around in our bodies and sticks them together in all sorts of combinations. The combinations that do not work separate and continue with their random attachment however, the combinations that do work produce things like ATP synthase. Through this process nature has been able to make a nanomachine that is 95% efficient, which is a feat that man has not been able to accomplish yet. This is all because nature does not try to work around the forces it uses them at its advantage.

Growing cells in culture to take advantage of their internal chemical synthesis machinery can be considered a form of nanotechnology but this machinery has also been manipulated outside of living cells.

5

Self-Assembly of Nanoparticles

Self-assembly is a phenomenon where the components of a system assemble themselves spontaneously via an interaction to form a larger functional unit. This spontaneous organisation can be due to direct specific interaction and/or indirectly through their environment. Due to the increasing technological advancements, the study of materials in the nanometre scale is becoming more important. The ability to assemble nanoparticles into well-defined configuration in space is crucial to the development of electronic devices that are small but can contain plenty of information. The spatial arrangements of these self-assembled nanoparticles can be potentially used to build increasingly complex structures leading to a wide variety of materials that can be used for different purposes.

At the molecular level, intermolecular force hold the spontaneous gathering of molecules into a well-defined and stable structure together. In chemical solutions, self-assembly is an outcome of random motion of molecules and the affinity of their binding sites for one another. In the area of nanotechnology, developing a simple, efficient method to organise molecules and molecular clusters into precise, pre-determined structure is crucial.

The study of self-assembly of nanoparticles is a new area in nanotechnology research. Scientists who are focusing their study on nanotechnology recognised that some properties of atoms and molecules enable them to arrange themselves into patterns. There are a variety of applications where the self-assembly of nanoparticles can be useful.

For example, building sensors to detect chemical and biological molecules. In addition, it can also be used on creating computer chips with smaller component sizes, which can then allow more computing power to be stored on a chip. Recently, researchers were able to watch nanoparticles to self-assembly for the very first time in real-time. This research was done by scientists at the U.S. Department of Energy's (DOE) Argonne National Laboratory. They had to use a transmission electron microscope (TEM) located at the Centre for Nanoscale Materials to capture the quick movement of the nanoparticles into a self-assembled system.

Thermodynamics

Self-assembly is an equilibrium process where the assembled components are in equilibrium with the individual components. Self-assembly is driven by the minimization of Gibbs free energy. The minimization of Gibbs free energy is attained by the minimization of repulsive and the maximization of attractive molecular interactions. In addition, the lower free energy is usually a result of a weaker intermolecular force between self-assembled moieties and is essentially enthalpic in nature.

The thermodynamics of the self-assembly process can be represented by a simple Gibbs free energy equation:

$$\Delta G_{SA} = \Delta H_{SA} - T\Delta S_{SA}$$

where if ΔG_{SA} is negative, self-assembly is a spontaneous process. ΔH_{SA} is the enthalpy change of the process and is largely determined by the potential energy/intermolecular forces between the assembling entities. ΔS_{SA} is the change in entropy associated with the formation of the ordered arrangement. In general, the organisation is accompanied by a decrease in entropy and in order for the assembly to be spontaneous the enthalpy term must be negative and in excess of the entropy term. This equation shows that as the value of $T\Delta S_{SA}$ approaches the value of and above a critical temperature, the self-assembly process will become progressively less likely to occur and spontaneous self-assembly will not happen.

The self-assembly is governed by the normal processes of nucleation and growth. Small assemblies are formed because of their increased lifetime as the attractive interactions between the components lower the Gibbs free energy. As the assembly grows, the Gibbs free energy continues to decrease until the assembly becomes stable enough to last for a long period of time. The necessity of the self-assembly to be an equilibrium process is defined by the organisation of the structure which requires non-ideal arrangements to be formed before the lowest energy configuration is found.

Defects

In a self-assembled structure a reasonable concentration of defects is present. Defects can be defined thermodynamically. In most cases, the thermodynamic driving force for self-assembly is provided by weak intermolecular interactions and is usually of the same order of magnitude as the entropy term. In order for a self-assembling system to reach the minimum free energy configuration, there has to be enough thermal energy to allow the mass transport of the self-assembling molecules. For defect formation, the free energy of single defect formation is given by:

$$\Delta G_{DF} = \Delta H_{DF} - T\Delta S_{DF}$$

The enthalpy term, ΔH_{DF} does not necessarily reflect the intermolecular forces between the molecules, it is the energy cost associated with disrupting the pattern and may be thought of as a region where optimum arrangement does not occur and the reduction of enthalpy associated with ideal self-assembly did not occur. An example of this can be seen in a system of hexagonally packed cylinders where defect regions of lamellar structure exist.

If ΔG_{DF} is negative, there will be a finite number of defects in the system and the concentration will be given by:

$$\frac{N}{N_0} = exp(\frac{-\Delta E_{act}}{RT})$$

N is the number of defects in a matrix of N_0 self-assembled particles or features and ΔE_{act} is the activation energy of defect formation. The activation energy, ΔE_{act} , should not be confused with ΔH_{DF} . The activation energy represents the energy difference between the initial ideally arranges state and a transition state towards the defective structure. At low defect concentrations, defect formation is entropy driven until a critical concentration of defects allows the activation energy term to compensate for entropy. There is usually an equilibrium defect density indicated at the minimum free energy. The activation energy for defect formation increases this equilibrium defect density.

Particle Interaction

Intermolecular forces govern the particle interaction in self-assembled systems. The forces tend to be intermolecular in type rather than ionic or covalent because ionic or covalent bonds will "lock" the assembly into non-equilibrium structures. The types intermolecular forces seen in self-assembly processes are van der Waals, hydrogen bonds, and weak polar forces, just to name a few. In self-assembly, regular structural arrangements are frequently observed, therefore there must be a balance of attractive and repulsive between molecules otherwise an equilibrium distance will not exist between the particles. The repulsive forces can be electron cloud-electron cloud overlap or electrostatic repulsion.

Processing

The processes by which nanoparticles self assemble are widespread and important. Understanding why and how self assembly occurs is key in reproducing and optimizing results. Typically, nanoparticles will self assemble for one or both of two reasons: molecular interactions and external direction. Recently, researchers are more interested in the latter of these two reasons, and are able to apply external directing to a large variety of systems.

Self Assembly by Molecular Interactions

Nanoparticles have the ability to assemble chemically through covalent or noncovalent interactions with their capping ligand. The terminal functional group(s) on the particle are known as capping ligands. As these ligands tend to be complex and sophisticated, self-assembly can provide a simpler pathway for nanoparticle organisation by synthesizing efficient functional groups.

Intermolecular Forces

Nanoparticles can self-assemble as a result of their intermolecular forces. As systems look to minimize their free energy, self assembly is one option for the system to achieve its lowest free energy thermodynamically. Nanoparticles can be programmed to self assemble by changing the functionality of their side groups, taking advantage of weak and specific intermolecular forces to spontaneously order the particles. These direct interparticle interactions can be typical intermolecular forces such as hydrogen bonding or Van der Waals forces, but can also be internal characteristics, such as hydrophobicity or hydrophilicity. For example, lipophilic nanoparticles have the tendency to self-assemble and form crystals as solvents are evaporated. While these aggregations are based on intermolecular forces, external factors such as temperature and pH also play a role in spontaneous self-assembly.

Hamaker Interaction

As nanoparticle interactions take place on a nanoscale, the particle interactions must be scaled similarly. Hamaker interactions take into account the polarization characteristics of a large number of nearby particles and the effects they have on each other. Hamaker interactions sum all of the forces between all particles and the solvent(s) involved in the system. While Hamaker theory generally describes a macroscopic system, the vast number of nanoparticles in a self assembling system allows the term to be applicable. Hamaker constants for nanoparticles are calculated using Lifshitz theory, and can often be found in literature.

Hamaker Constants for Nanoparticles in Water

Material	A_{131}
Fe_3O_4	22
$-Fe_2O_3$	26
a-Fe_2O_3	29
Ag	33
Au	45

All values reported in zJ

Externally Directed Self Assembly

The natural ability of nanoparticles to self assemble can be replicated in systems that do not intrinsically self assemble. Directed self assembly (DSA) attempts to mimic the chemical properties of self-assembling systems, while simultaneously controlling the thermodynamic system to maximize self assembly.

Electric and Magnetic Fields

External fields are the most common directors of self-assembly. Electric and magnetic fields allow induced interactions to align the particles. The fields take advantage of the polarizability of the nanoparticle and its functional groups. When these field-induced interactions overcome random Brownian motion, particles join to form chains and then assemble. At more modest field strengths, ordered crystal structures are established due to the induced dipole interactions. Electric and magnetic field direction requires a constant balance between thermal energy and interaction energies.

Flow Fields

Macroscopic Viscous Flow

Macroscopic viscous flow fields can direct self assembly of a random solution of particles into ordered crystals. However, the assembled particles tend to disassemble when the flow is stopped or removed. Shear flows are useful for jammed suspensions or random close packing. As these systems begin in nonequilibrium, flow fields are useful in that they help the system relax towards ordered equilibrium. Flow fields are also useful when dealing with complex matrices that themselves have rheological behaviour. Flow can induce anisotropic viseoelastic stresses, which helps to overcome the matrix and cause self assembly.

Large Amplitude Oscillatory Shear (LAOS)

Large amplitude oscillatory shear (LAOS) is most effective for particles that are 100 nm-1 μm in size. Hard and soft shears can order in steady shear. However, this assembly strongly relies on particle volume fraction, particle interaction potentials, polydisterity, and shear rate and strain. The large amount of directing factors can cause complications in directing self assembly by LAOS. Diblock copolymer micells have recently been studied in regard to structuring nanoparticles in bulk.

Combination of Fields

The most effective self assembly director is a combination of fields. If the fields and conditions are optimized, self assembly can be permanent and complete. When a field combination is used with nanoparticles that are tailored to be intrinsically responsive, the most complete assembly is observed. Combinations of fields allow the benefits of self assembly, such as scalability and simplicity, to be maintained while being able to control orientation and structure formation. Field combinations possess the greatest potential for future directed self assembly work.

Interfaces

Solid Interfaces

Nano-particles can self assemble on solid surfaces after applying external forces (like magnetic, electric, and flow) as mentioned in the above section. Templates made of microstructures like carbon nanotubes or block polymers can also be used to assist in self assembly; they cause directed self assembly (DSA) in which active sites are embedded to selectively induce nanoparticle deposition. Such templates are considered as any object onto which different particles can be arranged into a structure with a morphology similar to that of the template. Carbon nanotubes (microstructures), single molecules, or block copolymers are common templates.

Nanoparticles are often shown to self assemble within distances of nanometres and micrometres, but block copolymer templates can be used to form well-defined self-assemblies over macroscopic distances. By incorporating active sites to the surfaces of nanotubes and polymers, the functionalization of these templates can be transformed to favour self assembly of specified nanoparticles.

Liquid Interfaces

Pickering and Ramsden explained the idea of pickering emulsions when experimenting with paraffin-water emulsions with solid particles like iron oxide and silicon dioxide. They observed that the micron-sized colloids generated a resistant film at the interface between the two immiscible phases, inhibiting the coalescence of the emulsion drops. These Pickering emulsions, as shown in the figure below, are formed from the self-assembly of colloidal particles in two-part liquid systems, such as oil-water systems. The desorption energy, which is directly related to the stability of emulsions depends on the particle size, particle-particle interaction and, of course, particle-water and particle-oil interactions.

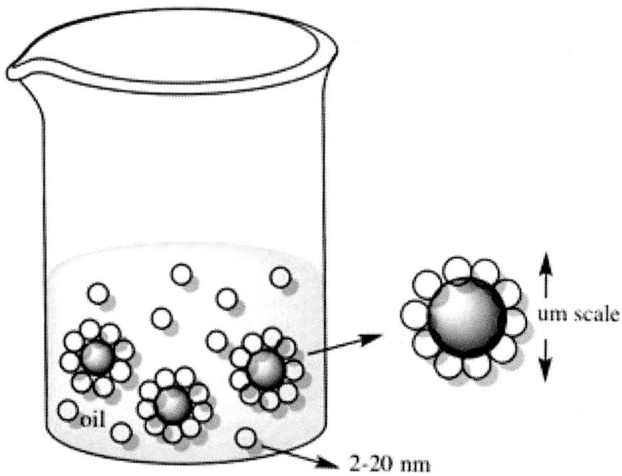

***Figure:** Self-assembly of solid nanoparticles at oil-water interface.*

A decrease in total free energy was observed to be a result from the assembly of nanoparticles at an oil/water (O/W) interface. This is shown in the following equation in which a particle with radius r at an interface between oil (O) and water (W) results in a decrease of the initial interfacial energy E_0 to E_1; this difference in energy is ΔE_1.

$$E_0 - E_1 = -\frac{\pi r^2}{\gamma_{O/W}}[\gamma_{O/W} - (\gamma_{P/W} - \gamma_{P/O})]^2$$

At a fixed $\gamma_{P/O}$, $\gamma_{P/W}$, and $\gamma_{O/W}$, the equation shows that the stability of a particle assembly is determined by the radius square. When moving to the interface, particles reduce the unfavourable contact between the immiscible fluids and decrease the interfacial energy. The decrease in total free energy for microscopic particles is much larger than that of thermal energy; this results in an effective confinement of large colloids to the interface. They are irreversibly bound to the interface. Nanoscopic particles are restricted to the interface by an energy reduction comparable to thermal energy. Thus, nanoparticles are easily displaced from the interface. A constant particle exchange then occurs at the interface; the rate of this exchange depends on particle size. The thermally activated escape of small particles occurs more often than larger particles. For the equilibrium state of assembly, the total gain in free energy is smaller for smaller particles. Thus, large nanoparticles assemblies are more stable. This size dependence allows nanoparticles to self-assemble at the interface to attain its equilibrium structure. Micrometre- size colloids, on the other hand, may be confined in a non-equilibrium state. The interfacial tension and the wettability of a particle surface affect the desorption energy. The contact angle θ between the solid and the oil/water interface determines its wettability. As shown in the figure below, a contact angle θ greater than 90° favours a water-in-oil emersion while a contact angle θ less than 90° favours oil-in-water emulsion. These contact angles affect the stability of the emulsion.

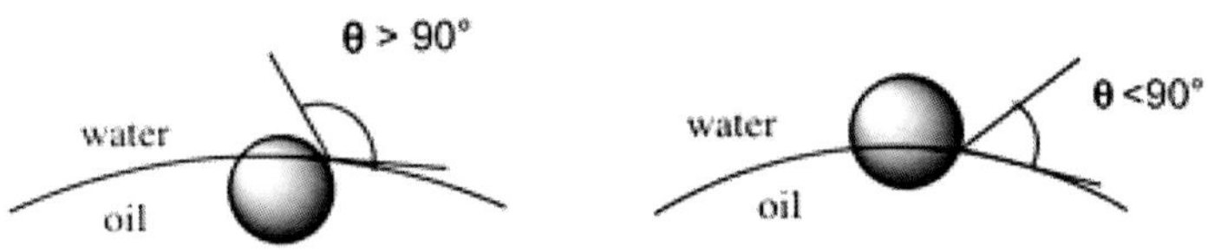

***Figure:** Effects of contact angle on wettability.*

A maximum desorption energy peak is observed at a contact angle of 90°. When the contact angle is greater than or less than this point, the desorption energy gradually decreases; thus, the stability of the emulsion decreases as well.

Applications

Material that consists with nano particles is called “nanostructured materials”. The phase of “nanostructured material” implies two important ideas: i) that at lease some of the heterogeneity in materials is determined by the size range of nanostructures (~1-100 nm), and ii) these nanostructures might be synthesized and distributed (or organised), at least in part. The study of self-assembly nanoparticles is important to understand the interaction between single particle in terms of applying them into different applications.

Electronics

Self-assembly of nanoscale structures from functional nanoparticles has provided a powerful path to developing small and powerful electronic components. In the recent decade, it has been confirmed that organic compounds can be conductors or semiconductors. In addition to insulator, one of the opportunities in nanomaterials science is to use organic synthesis and molecular design to make electronically usefully structures starting with organic molecules. The difficulty of applied nanostructure material is nanoscale objects have always been difficult to work with because they cannot be characterized by molecular techniques and they are too small to see optically. 2D self assembly monodisperse particle colloids has a strong potential in dense magnetic storage media. Each colloid particle has the ability to store information as known as binary number 0 and 1 after applying it to a strong magnetic field. In the meantime, it requires a nanoscale sensor or detector in order to selectively choose the colloid particle.

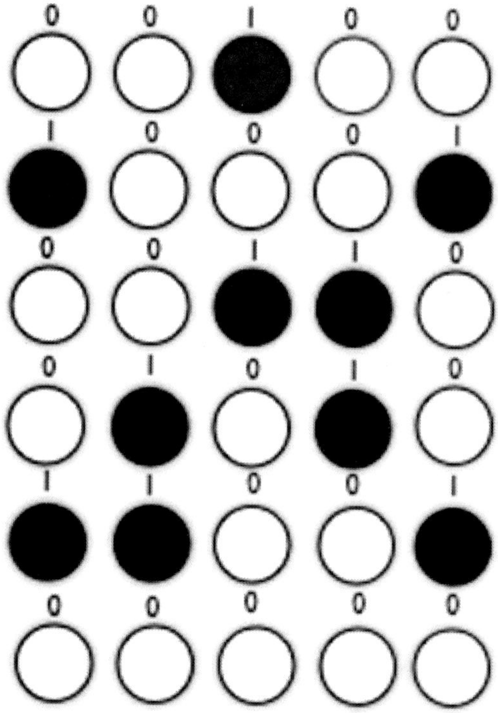

Figure: *Model of multidimensional array of nano-particles. Particle could have two spins, spin up or down. Based on the spin directions, nano-particles will be able to store 0 and 1. Therefore, nanostructural material has a great potential to be new material that can be used widely electronic devices.*

Biological Applications

Drug Delivery

Block copolymers offer the ability to self-assemble into uniform, nanosized micelles and accumulate in tumors via the enhanced permeability and retention effect. Polymer composition can be chosen to control the micelle size and compatibility with the drug of choice.

The challenges of this application are the difficulty of reproducing or controlling the size of self-assembly nano micelle, preparing predictable size-distribution, and the stability of the micelle with high drug load content.

Icosahedral Twins

An icosahedral-twin is a twenty-face cluster made of ten interlinked dual-tetrahedron (bowtie) crystals, typically joined along triangular (e.g. cubic-111) faces having three-fold symmetry. One can think of their formation as a kind of nano-scale self-assembly.

A variety of nanostructures (e.g. condensing argon, metal atoms, and virus capsids) assume icosahedral form on size scales where surface forces eclipse those from the bulk. A twinned-form of these nanostructures is sometimes found to occur e.g. in face-centred-cubic (FCC) metal-atom clusters larger than 10 nm in diameter. This may occur when the building-blocks beneath each of the 20 facets of an initially icosahedral cluster make the case for conversion to a translationally symmetric crystalline form.

Causes

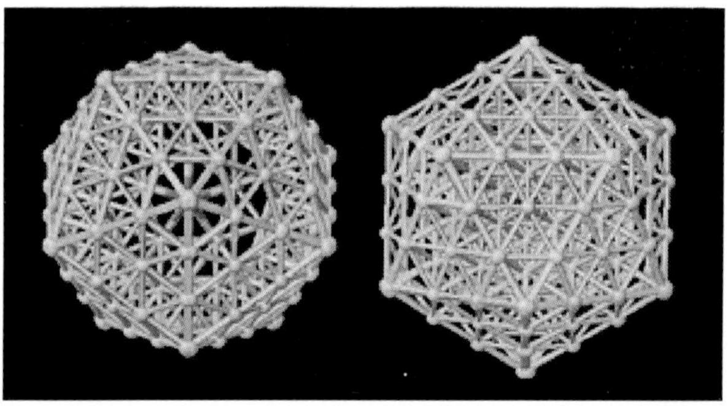

Figure: *FCC icotwin model down 5-fold & 3-fold zones.*

When interatom-bonding doesn't have strong directional preferences, it's not unusual for atoms to gravitate toward a kissing-number of 12 nearest-neighbours. The three most symmetric ways to do this are by icosahedral clustering, or by crystalline face-centred-cubic (cuboctahedral) and/or hexagonal (tri-orthobicupolar) close-packing.

Icosahedral arrangements, perhaps because of their slightly smaller surface area, may be preferred for small clusters e.g. noble gas and metal atoms in condensed phases (both liquid and solid). However, the Achilles-heel for icosahedral clustering about a single point is that it cannot fill space over large distances in a way that is translationally ordered.

Hence bulk atoms (i.e. sufficiently large clusters) generally revert to one of the crystalline close-packing configurations instead.

In other words, when icosahedral clusters get sufficiently large, the bulk-atom vote wins out over the surface-atom vote, and the atoms beneath each of the 20 facets adopt a face-centred-cubic pyramidal arrangement with tetrahedral (111) facets. Thus icosahedral twins are born, with a certain amount of strain along the interfacial (111) planes.

Ubiquity

Icosahedral-twinning has been seen in face-centred-cubic metal-nanoparticles that have nucleated:

(i) by evaporation onto surfaces,

(ii) out of solution, and

(iii) by reduction in a polymer matrix.

Icosahedral (or decahedral) twinning seems to be the result when clusters growing outward from an icoashedral-seed get larger than about 10[nm] in diameter.

Quasicrystals are *un-twinned* structures with long range rotational but not translational periodicity, that some initially tried to explain away as icosahedral twinning.

Quasi-crystals let non-space-filling coordination persist to larger size scales. However, they generally form only when the compositional makeup (e.g. of two dissimilar metals like Ti and Mn) serves as an antagonist to formation of one of the more common close-packed space-filling but twinned crystalline forms.

Application

Face-centred-cubic noble-metal atom-clusters are important chemical-reaction nano-catalysts. One example of this is the platinum used in automobile catalytic converters. Icosahedral twinning makes it possible to cover the entire surface of a nanoparticle with {111} facets, in cases where those particular atomic-facets show favourable catalytic-activity.

Detection

Electron diffraction and high-resolution transmission electron microscope (TEM) imaging are two methods for identifying the icosahedral-twin structure of *individual* clusters. Digital darkfield analysis of lattice-fringe images shows promise for recognition of icosahedral twinning from most of the randomly oriented clusters in a microscope-image field of view.

Impact of Nanotechnology

The impact of nanotechnology extends from its medical, ethical, mental, legal and environmental applications, to fields such as engineering, biology, chemistry, computing, materials science, and communications.

Major benefits of nanotechnology include improved manufacturing methods, water purification systems, energy systems, physical enhancement, nanomedicine, better food production methods, nutrition and large scale infrastructure auto-fabrication. Nanotechnology's reduced size may allow for automation of tasks which were previously inaccessible due to physical restrictions, which in turn may reduce labor, land, or maintenance requirements placed on humans.

Potential risks include environmental, health, and safety issues; transitional effects such as displacement of traditional industries as the products of nanotechnology become dominant, which are of concern to privacy rights advocates. These may be particularly important if potential negative effects of nanoparticles are overlooked.

Whether nanotechnology merits special government regulation is a controversial issue. Regulatory bodies such as the United States Environmental Protection Agency and the Health & Consumer Protection Directorate of the European Commission have started dealing with the potential risks of nanoparticles. The organic food sector has been the first to act with the regulated exclusion of engineered nanoparticles from certified organic produce, firstly in Australia and the UK, and more recently in Canada, as well as for all food certified to Demeter International standards.

Overview

Potential Risks: Potential risks of nanotechnology can broadly be grouped into four areas:

- Health issues - the effects of nanomaterials on human biology
- Environmental issues - the effects of nanomaterials on the environment
- Societal issues - the effects that the availability of nanotechnological devices will have on politics and human interaction
- Speculative issues - the specific risks associated with the speculative vision of molecular nanotechnology, such as "Grey goo"

Health and Safety Impact from Nanoparticles

The presence of nanomaterials (materials that contain nanoparticles) is not in itself a threat. It is only certain aspects that can make them risky, in particular their mobility and their increased reactivity. Only if certain properties of certain nanoparticles were harmful to living beings or the environment would we be faced with a genuine hazard. In this case it can be called nanopollution.

In addressing the health and environmental impact of nanomaterials we need to differentiate between two types of nanostructures:

(1) Nanocomposites, nanostructured surfaces and nanocomponents (electronic, optical, sensors etc.), where nanoscale particles are incorporated into a substance, material or device ("fixed" nanoparticles); and

(2) "free" nanoparticles, where at some stage in production or use individual nanoparticles of a substance are present.

These free nanoparticles could be nanoscale species of elements, or simple compounds, but also complex compounds where for instance a nanoparticle of a particular element is coated with another substance ("coated" nanoparticle or "core-shell" nanoparticle).

There seems to be consensus that, although one should be aware of materials containing fixed nanoparticles, the immediate concern is with free nanoparticles.

Nanoparticles are very different from their everyday counterparts, so their adverse effects cannot be derived from the known toxicity of the macro-sized material. This poses significant issues for addressing the health and environmental impact of free nanoparticles.

To complicate things further, in talking about nanoparticles it is important that a powder or liquid containing nanoparticles almost never be monodisperse [1], but contain instead a range of particle sizes. This complicates the experimental analysis as larger nanoparticles might have different properties from smaller ones. Also, nanoparticles show a tendency to aggregate, and such aggregates often behave differently from individual nanoparticles.

The National Institute for Occupational Safety and Health has conducted initial research on how nanoparticles interact with the body's systems and how workers might be exposed to nano-sized particles in the manufacturing or industrial use of nanomaterials. NIOSH currently offers interim guidelines for working with nanomaterials consistent with the best scientific knowledge. At The National Personal Protective Technology Laboratory of NIOSH, studies investigating the filter penetration of nanoparticles on NIOSH-certified and EU marked repirators, as well as non-certified dust masks have been conducted. These studies found that the most penetrating particle size range was between 30 and 100 nanometres, and leak size was the largest factor in the number of nanoparticles found inside the respirators of the test dummies.

In "The Consumer Product Safety Commission and Nanotechnology," E. Marla Felcher suggests that the Consumer Product Safety Commission, which is charged with protecting the public against unreasonable risks of injury or death associated with consumer products, is ill-equipped to oversee the safety of complex, high-tech products made using nanotechnology.

Longer-term concerns centre on the impact that new technologies will have for society at large, and whether these could possibly lead to either a post-scarcity economy, or alternatively exacerbate the wealth gap between developed and developing nations. The effects of nanotechnology on the society as a whole, on human health and the environment, on trade, on security, on food systems and even on the definition of "human", have not been characterized or politicized.

Health Issues

The health impact of nanotechnology are the possible effects that the use of nanotechnological materials and devices will have on human health. As nanotechnology is an emerging field, there is great debate regarding to what extent nanotechnology will benefit or pose risks for human health. Nanotechnology's health impact can be split into two aspects: the potential for nanotechnological innovations to have medical applications to cure disease, and the potential health hazards posed by exposure to nanomaterials.

Nanotoxicology is the field which studies potential health risks of nanomaterials. The extremely small size of nanomaterials means that they are much more readily taken up by the human body than larger sized particles. How these nanoparticles behave inside the organism is one of the significant issues that needs to be resolved. The behaviour of nanoparticles is a function of their size, shape and surface reactivity with the surrounding tissue. Apart from what happens if non-degradable or slowly degradable nanoparticles accumulate in organs, another concern is their potential interaction with biological processes inside the body: because of their large surface, nanoparticles on exposure to tissue and fluids will immediately adsorb onto their surface some of the macromolecules they encounter. The large number of variables influencing toxicity means that it is difficult to generalise about health risks associated with exposure to nanomaterials – each new nanomaterial must be assessed individually and all material properties must be taken into account. Health and environmental issues combine in the workplace of companies engaged in producing or using nanomaterials and in the laboratories engaged in nanoscience and nanotechnology research. It is safe to say that current workplace exposure standards for dusts cannot be applied directly to nanoparticle dusts.

Nanomedicine is the medical application of nanotechnology. The approaches to nanomedicine range from the medical use of nanomaterials, to nanoelectronic biosensors, and even possible future applications of molecular nanotechnology. Nanomedicine seeks to deliver a valuable set of research tools and clinically helpful devices in the near future. The National Nanotechnology Initiative expects new commercial applications in the pharmaceutical industry that may include advanced drug delivery systems, new therapies, and in vivo imaging. Neuro-electronic interfaces and other nanoelectronics-based sensors are another active goal of research. Further down the line, the speculative field of molecular nanotechnology believes that cell repair machines could revolutionize medicine and the medical field.

Environmental Issues

Nanopollution is a generic name for all waste generated by nanodevices or during the nanomaterials manufacturing process. This kind of waste may be very dangerous because of its size. It can float in the air and might easily penetrate animal and plant cells causing unknown effects. Most human-made nanoparticles do not appear in nature, so living organisms may not have appropriate means to deal with nanowaste.

To properly assess the health hazards of engineered nanoparticles the whole life cycle of these particles needs to be evaluated, including their fabrication, storage and distribution, application and potential abuse, and disposal. The impact on humans or the environment may vary at different stages of the life cycle. Environmental assessment is justified as nanoparticles present novel environmental impacts. Scrinis raises concerns about nano-pollution, and argues that it is not currently possible to "precisely predict or control the ecological impacts of the release of these nano-products into the environment."

On the other hand, some possible future applications of nanotechnology have the potential to benefit the environment. Nanofiltration, based on the use of membranes with extremely small pores smaller than 10 nm (perhaps composed of nanotubes) are suitable for a mechanical filtration for the removal of ions or the separation of different fluids. Furthermore, magnetic nanoparticles offer an effective and reliable method to remove heavy metal contaminants from waste water. Using nanoscale particles increases the efficiency to absorb the contaminants and is comparatively inexpensive compared to traditional precipitation and filtration methods.

Furthermore, nanotechnology could potentially have a great impact on clean energy production. Research is underway to use nanomaterials for purposes including more efficient solar cells, practical fuel cells, and environmentally friendly batteries.

Regulation Issues

Significant debate exists relating to the question of whether nanotechnology or nanotechnology-based products merit special government regulation. This debate is related to the circumstances in which it is necessary and appropriate to assess new substances prior to their release into the market, community and environment.

Regulatory bodies such as the United States Environmental Protection Agency and the Food and Drug Administration in the U.S. or the Health & Consumer Protection Directorate of the European Commission have started dealing with the potential risks posed by nanoparticles. So far, neither engineered nanoparticles nor the products and materials that contain them are subject to any special regulation regarding production, handling or labelling. The Material Safety Data Sheet that must be issued for some materials often does not differentiate between bulk and nanoscale size of the material in question and even when it does these MSDS are advisory only.

Limited nanotechnology labelling and regulation may exacerbate potential human and environmental health and safety issues associated with nanotechnology. It has been argued that the development of comprehensive regulation of nanotechnology will be vital to ensure that the potential risks associated with the research and commercial application of nanotechnology do not overshadow its potential benefits. Regulation may also be required to meet community expectations about responsible development of nanotechnology, as well as ensuring that public interests are included in shaping the development of nanotechnology.

Societal Impact

Beyond the toxicity risks to human health and the environment which are associated with first-generation nanomaterials, nanotechnology has broader societal impact and poses broader social challenges. Social scientists have suggested that nanotechnology's social issues should be understood and assessed not simply as "downstream" risks or impacts.

Rather, the challenges should be factored into "upstream" research and decision making in order to ensure technology development that meets social objectives.

Many social scientists and organisations in civil society suggest that technology assessment and governance should also involve public participation.

The last few years has seen a gold rush to claim patents at the nanoscale. Over 800 nano-related patents were granted in 2003, and the numbers are increasing year to year. Corporations are already taking out broad-ranging patents on nanoscale discoveries and inventions. For example, two corporations, NEC and IBM, hold the basic patents on carbon nanotubes, one of the current cornerstones of nanotechnology. Carbon nanotubes have a wide range of uses, and look set to become crucial to several industries from electronics and computers, to strengthened materials to drug delivery and diagnostics. Carbon nanotubes are poised to become a major traded commodity with the potential to replace major conventional raw materials. However, as their use expands, anyone seeking to (legally) manufacture or sell carbon nanotubes, no matter what the application, must first buy a license from NEC or IBM. [2] [3]

Potential Benefits and Risks for Developing Countries

Nanotechnologies may provide new solutions for the millions of people in developing countries who lack access to basic services, such as safe water, reliable energy, health care, and education. The United Nations has set Millennium Development Goals for meeting these needs. The 2004 UN Task Force on Science, Technology and Innovation noted that some of the advantages of nanotechnology include production using little labor, land, or maintenance, high productivity, low cost, and modest requirements for materials and energy.

Potential opportunities of nanotechnologies to help address critical international development priorities include improved water purification systems, energy systems, medicine and pharmaceuticals, food production and nutrition, and information and communications technologies. Nanotechnologies are already incorporated in products that are on the market. Other nanotechnologies are still in the research phase, while others are concepts that are years or decades away from development.

Protection of the environment, human health and worker safety in developing countries often suffers from a combination of factors that can include but are not limited to lack of robust environmental, human health, and worker safety regulations; poorly or unenforced regulation which is linked to a lack of physical (e.g., equipment) and human capacity (i.e., properly trained regulatory staff). Often, these nations require assistance, particularly financial assistance, to develop the scientific and institutional capacity to adequately assess and manage risks, including the necessary infrastructure such as laboratories and technology for detection.

However, concerns are frequently raised that the claimed benefits of nanotechnology will not be evenly distributed, and that any benefits (including technical and/or economic) associated with nanotechnology will only reach affluent nations. The majority of nanotechnology research and development - and patents for nanomaterials and products - is concentrated in developed countries (including the United States, Japan, Germany, Canada and France). In addition, most patents related to nanotechnology are concentrated amongst few multinational corporations, including IBM, Micron Technologies, Advanced Micro Devices and Intel. This has led to fears that it will be unlikely that developing countries will have access to the infrastructure, funding and human resources required to support nanotechnology research and development, and that this is likely to exacerbate such inequalities.

Producers in developing countries could also be disadvantaged by the replacement of natural products (including rubber, cotton, coffee and tea) by developments in nanotechnology. These natural products are important export crops for developing countries, and many farmers' livelihoods depend on them. It has been argued that their substitution with industrial nano-products could negatively affect the economies of developing countries, that have traditionally relied on these export crops.

Effects on Laborers

Ray Kurzweil has speculated in *The Singularity is Near* that people who work in unskilled labor jobs for a livelihood may become the first human workers to be displaced by the constant use of nanotechnology in the workplace, noting that layoffs often affect the jobs based around the lowest technology level before attacking jobs with the highest technology level possible. It has been noted that every major economic era has stimulated a global revolution both in the kinds of jobs that are available to people and the kind of training they need to achieve these jobs, and there is concern that the world's educational systems have lagged behind in preparing students for the "Nanotech Age".

It has also been speculated that nanotechnology may give rise to nanofactories which may have superior capabilities to conventional factories due to their small carbon and physical footprint on the global and regional environment. The miniaturization and transformation of the multi-acre conventional factory into the nanofactory may not interfere with their ability to deliver a high quality product; the product may be of even greater quality due to the lack of human errors in the production stages. Nanofactory systems may use precise atomic precisioning and contribute to making superior quality products that the "bulk chemistry" method used in 20th century and early 21st currently cannot produce. These advances might shift the computerized workforce in an even more complex direction, requiring skills in genetics, nanotechnology, and robotics.

Impact of Molecular Nanotechnology

Molecular nanotechnology is a speculative subfield of nanotechnology regarding the possibility of engineering molecular assemblers, machines which could re-order matter at a molecular or atomic scale. Regarding the risks from molecular manufacturing, an often cited worst-case scenario is "grey goo", a hypothetical substance into which the surface of the earth might be transformed by self-replicating nanobots running amok. This concept has been analyzed by Freitas in "Some Limits to Global Ecophagy by Biovorous Nanoreplicators, with Public Policy Recommendations" With the advent of nano-biotech, a different scenario called green goo has been forwarded. Here, the malignant substance is not nanobots but rather self-replicating organisms engineered through nanotechnology.

According to the Centre for Responsible Nanotechnology: Molecular manufacturing allows the cheap creation of incredibly powerful devices and products. How many of these products will we want? What environmental damage will they do? The range of possible damage is vast, from personal low-flying supersonic aircraft injuring large numbers of animals to collection of solar energy on a sufficiently large scale to modify the planet's albedo and directly affect the environment. Stronger materials will allow the creation of much larger machines, capable of excavating or otherwise destroying large areas of the planet at a greatly accelerated pace.

It is too early to tell whether there will be economic incentive to do this. However, given the large number of activities and purposes that would damage the environment if taken to extremes, and the ease of taking them to extremes with molecular manufacturing, it seems likely that this problem is worth worrying about. Some forms of damage can result from an aggregate of individual actions, each almost harmless by itself. Such damage is quite hard to prevent by persuasion, and laws frequently don't work either; centralized restriction on the technology itself may be a necessary part of the solution.

Finally, the extreme compactness of nanomanufactured machinery will tempt the use of very small products, which can easily turn into nano-litter that will be hard to clean up and may cause health problems.

The site list numerous other risks and benefits.

Nanocrystalline Silicon

Nanocrystalline silicon (nc-Si), sometimes also known as microcrystalline silicon (μc-Si), is a form of porous silicon. It is an allotropic form of silicon with paracrystalline structure—is similar to amorphous silicon (a-Si), in that it has an amorphous phase. Where they differ, however, is that nc-Si has small grains of crystalline silicon within the amorphous phase. This is in contrast to polycrystalline silicon (poly-Si) which consists solely of crystalline silicon grains, separated by grain boundaries.

The difference comes solely from the grain size of the crystalline grains. Most materials with grains in the micrometre range are actually fine-grained polysilicon, so nanocrystalline silicon is a better term. The term Nanocrystalline silicon refers to a range of materials around the transition region from amorphous to microcrystalline phase in the silicon thin film. The crystalline volume fraction (as measured from Raman spectroscopy) is another criterion to describe the materials in this transition zone.

nc-Si has many useful advantages over a-Si, one being that if grown properly it can have a higher electron mobility, due to the presence of the silicon crystallites. It also shows increased absorption in the red and infrared wavelengths, which make it an important material for use in a-Si solar cells. One of the most important advantages of nanocrystalline silicon, however, is that it has increased stability over a-Si, one of the reasons being because of its lower hydrogen concentration. Although it currently cannot attain the mobility that poly-Si can, it has the advantage over poly-Si that it is easier to fabricate, as it can be deposited using conventional low temperature a-Si deposition techniques, such as PECVD, as opposed to laser annealing or high temperature CVD processes, in the case of poly-Si.

Uses

The main application of this novel material is in the field of silicon thin film solar cells. As nc-Si has about the same bandgap as crystalline silicon, which is ~1.12 eV, it can be combined in thin layers with a-Si, creating a layered, multi-junction cell called a tandem cell. The top cell in a-Si absorbs the visible light and leaves the infrared part of the spectrum for the bottom cell in nanocrystalline Si.

A few companies are on the verge of commercializing silicon inks based on nanocrystalline silicon or on other silicon compounds. The semiconductor industry is also investigating the potential for nanocrystalline silicon, especially in the memory area.

Thin-Film Silicon

Nanocrystalline silicon and small-grained polycrystalline silicon are considered thin-film silicon.

Polycrystalline Silicon

Polycrystalline silicon, also called polysilicon, is a material consisting of small silicon crystals. It differs from single-crystal silicon, used for electronics and solar cells, and from amorphous silicon, used for thin film devices and solar cells.

Single Crystalline vs Polycrystalline Silicon

In single crystal silicon, the crystalline framework is homogenous, which can be recognised by an even external colouring. In single crystal silicon, also called monocrystal, the crystal lattice of the entire sample is continuous and unbroken with no grain boundaries. Large single crystals are exceedingly rare in nature and can also be difficult to produce in the laboratory. In contrast, in an amorphous structure the order in atomic positions is limited to short range.

DNA Nanotechnology

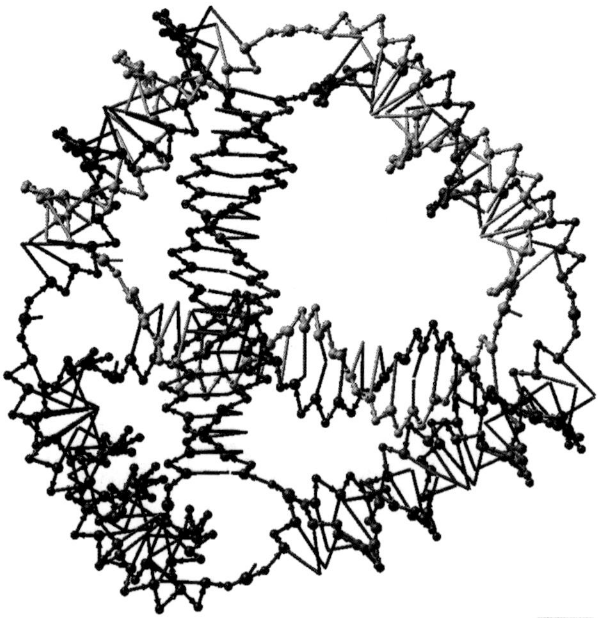

Figure: *DNA nanotechnology involves the creation of artificial, designed nanostructures out of nucleic acids, such as this DNA tetrahedron. Each edge of the tetrahedron is a 20 base pair DNA double helix, and each vertex is a three-arm junction. The 4 DNA strands that form the 4 tetrahedral faces are colour-coded.*

DNA nanotechnology is the design and manufacture of artificial nucleic acid structures for technological uses. In this field, nucleic acids are used as non-biological engineering materials for nanotechnology rather than as the carriers of genetic information in living cells. Researchers in the field have created static structures such as two- and three-dimensional crystal lattices, nanotubes, polyhedra, and arbitrary shapes, as well as functional devices such as molecular machines and DNA computers. The field is beginning to be used as a tool to solve basic science problems in structural biology and biophysics, including applications in crystallography and spectroscopy for protein structure determination. Potential applications in molecular scale electronics and nanomedicine are also being investigated.

The conceptual foundation for DNA nanotechnology was first laid out by Nadrian Seeman in the early 1980s, and the field began to attract widespread interest in the mid-2000s. This use of nucleic acids is enabled by their strict base pairing rules, which cause only portions of strands with complementary base sequences to bind together to form strong, rigid double helix structures. This allows for the rational design of base sequences that will selectively assemble to form complex target structures with precisely controlled nanoscale features. A number of assembly methods are used to make these structures, including tile-based structures that assemble from smaller structures, folding structures using the DNA origami method, and dynamically reconfigurable structures using strand displacement techniques. DNA is the dominant material used, but structures incorporating other nucleic acids such as RNA and peptide nucleic acid (PNA) have also been constructed, leading to the occasional use of the name nucleic acid nanotechnology to describe the field.

Properties of Nucleic Acids

Nanotechnology is often defined as the study of materials and devices with features on a scale below 100 nanometres. DNA nanotechnology, specifically, is an example of bottom-up molecular self-assembly, in which molecular components spontaneously organise into stable structures; the particular form of these structures is induced by the physical and chemical properties of the components selected by the designers. In DNA nanotechnology, the component materials are strands of nucleic acids such as DNA; these strands are often synthetic and are almost always used outside the context of a living cell. DNA is well-suited to nanoscale construction because the binding between two nucleic acid strands depends on simple base pairing rules which are well understood, and form the specific nanoscale structure of the nucleic acid double helix. These qualities make the assembly of nucleic acid structures easy to control through nucleic acid design. This property is absent in other materials used in nanotechnology, including proteins, for which protein design is very difficult, and nanoparticles, which lack the capability for specific assembly on their own.

The structure of a nucleic acid molecule consists of a sequence of nucleotides distinguished by which nucleobase they contain. In DNA, the four bases present are adenine (A), cytosine (C), guanine (G), and thymine (T). Nucleic acids have the property that two molecules will only bind to each other to form a double helix if the two sequences are complementary, meaning that they form matching sequences of base pairs, with A only binding to T, and C only to G. Because the formation of correctly matched base pairs is energetically favourable, nucleic acid strands are expected in most cases to bind to each other in the conformation that maximizes the number of correctly paired bases. The sequences of bases in a system of strands thus determine the pattern of binding and the overall structure in an easily controllable way. In DNA nanotechnology, the base sequences of strands are rationally designed by researchers so that the base pairing interactions cause the strands to assemble in the desired conformation.

Subfields

DNA nanotechnology is sometimes divided into two overlapping subfields: structural DNA nanotechnology and dynamic DNA nanotechnology. Structural DNA nanotechnology, sometimes abbreviated as SDN, focuses on synthesizing and characterizing nucleic acid complexes and materials that assemble into a static, equilibrium end state. On the other hand, dynamic DNA nanotechnology focuses on complexes with useful non-equilibrium behaviour such as the ability to reconfigure based on a chemical or physical stimulus. Some complexes, such as nucleic acid nanomechanical devices, combine features of both the structural and dynamic subfields.

The complexes constructed in structural DNA nanotechnology use topologically branched nucleic acid structures containing junctions. (In contrast, most biological DNA exists as an unbranched double helix.) One of the simplest branched structures is a four-arm junction that consists of four individual DNA strands, portions of which are complementary in a specific pattern. Unlike in natural Holliday junctions, each arm in the artificial immobile four-arm junction has a different base sequence, causing the junction point to be fixed at a certain position. Multiple junctions can be combined in the same complex, such as in the widely used double-crossover (DX) motif, which contains two parallel double helical domains with individual strands crossing between the domains at two crossover points. Each crossover point is itself topologically a four-arm junction, but is constrained to a single orientation, as opposed to the flexible single four-arm junction, providing a rigidity that makes the DX motif suitable as a structural building block for larger DNA complexes.

Dynamic DNA nanotechnology uses a mechanism called toehold-mediated strand displacement to allow the nucleic acid complexes to reconfigure in response to the addition of a new nucleic acid strand. In this reaction, the incoming strand binds to a single-stranded toehold region of a double-stranded complex, and then displaces one of the strands bound in the original complex through a branch migration process. The overall effect is that one of the strands in the complex is replaced with another one. In addition, reconfigurable structures and devices can be made using functional nucleic acids such as deoxyribozymes and ribozymes, which are capable of performing chemical reactions, and aptamers, which can bind to specific proteins or small molecules.

Design

DNA nanostructures must be rationally designed so that the individual nucleic acid strands will assemble into the desired structures. This process usually begins with the specification of a desired target structure or functionality. Then, the overall secondary structure of the target complex is determined, specifying the arrangement of nucleic acid strands within the structure, and which portions of those strands should be bound to each other. The last step is the primary structure design, which is the specification of the actual base sequences of each nucleic acid strand.

Structural Design

The first step in designing a nucleic acid nanostructure is to decide how a given structure should be represented by a specific arrangement of nucleic acid strands. This design step determines the secondary structure, or the positions of the base pairs that hold the individual strands together in the desired shape. Several approaches have been demonstrated:

- Tile-based structures. This approach breaks the target structure into smaller units with strong binding between the strands contained in each unit, and weaker interactions between the units. It is often used to make periodic lattices, but can also be used to implement algorithmic self-assembly, making them a platform for DNA computing. This was the dominant design strategy used from the mid-1990s until the mid-2000s, when the DNA origami methodology was developed.

- Folding structures. An alternative to the tile-based approach, folding approaches make the nanostructure from a single long strand. This long strand can either have a designed sequence that folds due to its interactions with itself, or it can be folded into the desired shape by using shorter, "staple" strands. This latter method is called DNA origami, which allows the creation of nanoscale two- and three-dimensional shapes.
- Dynamic assembly. This approach directly controls the kinetics of DNA self-assembly, specifying all of the intermediate steps in the reaction mechanism in addition to the final product. This is done using starting materials which adopt a hairpin structure; these then assemble into the final conformation in a cascade reaction, in a specific order. This approach has the advantage of proceeding isothermally, at a constant temperature. This is in contrast to the thermodynamic approaches, which require a thermal annealing step where a temperature change is required to trigger the assembly and favour proper formation of the desired structure.

Sequence Design

After any of the above approaches are used to design the secondary structure of a target complex, an actual sequence of nucleotides that will form into the desired structure must be devised. Nucleic acid design is the process of assigning a specific nucleic acid base sequence to each of a structure's constituent strands so that they will associate into a desired conformation. Most methods have the goal of designing sequences so that the target structure has the lowest energy, and is thus the most thermodynamically favourable, while incorrectly assembled structures have higher energies and are thus disfavoured. This is done either through simple, faster heuristic methods such as sequence symmetry minimization, or by using a full nearest-neighbour thermodynamic model, which is more accurate but slower and more computationally intensive. Geometric models are used to examine tertiary structure of the nanostructures and to ensure that the complexes are not overly strained.

Nucleic acid design has similar goals to protein design. In both, the sequence of monomers is designed to favour the desired target structure and to disfavour other structures. Nucleic acid design has the advantage of being much computationally easier than protein design, because the simple base pairing rules are sufficient to predict a structure's energetic favourability, and detailed information about the overall three-dimensional folding of the structure is not required. This allows the use of simple heuristic methods that yield experimentally robust designs. However, nucleic acid structures are less versatile than proteins in their functionality because of proteins' increased ability to fold into complex structures, as well as the limited chemical diversity of the four nucleotides as compared to the twenty proteinogenic amino acids.

Structural DNA Nanotechnology

Structural DNA nanotechnology, sometimes abbreviated as SDN, focuses on synthesizing and characterizing nucleic acid complexes and materials where the assembly has a static, equilibrium endpoint. The nucleic acid double helix has a robust, defined three-dimensional geometry that makes it possible to predict and design the structures of more complicated nucleic acid complexes. Many such structures have been created, including two- and three-dimensional structures, and periodic, aperiodic, and discrete structures.

Extended Lattices

Small nucleic acid complexes can be equipped with sticky ends and combined into larger two-dimensional periodic lattices containing a specific tessellated pattern of the individual molecular tiles. The earliest example of this used double-crossover (DX) complexes as the basic tiles, each containing four sticky ends designed with sequences that caused the DX units to combine into periodic two-dimensional flat sheets that are essentially rigid two-dimensional crystals of DNA. Two-dimensional arrays have been made from other motifs as well, including the Holliday junction rhombus lattice, and various DX-based arrays making use of a double-cohesion scheme. The top two images at right show examples of tile-based periodic lattices.

Two-dimensional arrays can be made to exhibit aperiodic structures whose assembly implements a specific algorithm, exhibiting one form of DNA computing. The DX tiles can have their sticky end sequences chosen so that they act as Wang tiles, allowing them to perform computation. A DX array whose assembly encodes an XOR operation has been demonstrated; this allows the DNA array to implement a cellular automaton that generates a fractal known as the Sierpinski gasket. The third image at right shows this type of array. Another system has the function of a binary counter, displaying a representation of increasing binary numbers as it grows. These results show that computation can be incorporated into the assembly of DNA arrays.

DX arrays have been made to form hollow nanotubes 4–20 nm in diameter, essentially two-dimensional lattices which curve back upon themselves. These DNA nanotubes are somewhat similar in size and shape to carbon nanotubes, and while they lack the electrical conductance of carbon nanotubes, DNA nanotubes are more easily modified and connected to other structures.

One of many schemes for constructing DNA nanotubes uses a lattice of curved DX tiles that curls around itself and closes into a tube. In an alternative method that allows the circumference to be specified in a simple, modular fashion using single-stranded tiles, the rigidity of the tube is an emergent property.

The creation of three-dimensional lattices out of DNA was the earliest goal of DNA nanotechnology, but this proved to be one of the most difficult to realise. Success using a motif based on the concept of tensegrity, a balance between tension and compression forces, was finally reported in 2009.

Discrete Structures

Researchers have synthesized a number of three-dimensional DNA complexes that each have the connectivity of a polyhedron, such as a cube or octahedron, meaning that the DNA duplexes trace the edges of a polyhedron with a DNA junction at each vertex. The earliest demonstrations of DNA polyhedra were very work-intensive, requiring multiple ligations and solid-phase synthesis steps to create catenated polyhedra.

Subsequent work yielded polyhedra whose synthesis was much easier. These include a DNA octahedron made from a long single strand designed to fold into the correct conformation, and a tetrahedron that can be produced from four DNA strands in a single step, pictured at the top of this article.

Nanostructures of arbitrary, non-regular shapes are usually made using the DNA origami method. These structures consist of a long, natural virus strand as a "scaffold", which is made to fold into the desired shape by computationally designed short "staple" strands. This method has the advantages of being easy to design, as the base sequence is predetermined by the scaffold strand sequence, and not requiring high strand purity and accurate stoichiometry, as most other DNA nanotechnology methods do.

DNA origami was first demonstrated for two-dimensional shapes, such as a smiley face and a coarse map of North America. Solid three-dimensional structures can be made by using parallel DNA helices arranged in a honeycomb pattern, and structures with two-dimensional faces can be made to fold into a hollow overall three-dimensional shape, akin to a cardboard box.

These can be programmed to open and reveal or release a molecular cargo in response to a stimulus, making them potentially useful as programmable molecular cages.

Templated Assembly

Nucleic acid structures can be made to incorporate molecules other than nucleic acids, sometimes called heteroelements, including proteins, metallic nanoparticles, quantum dots, and fullerenes. This allows the construction of materials and devices with a range of functionalities much greater than is possible with nucleic acids alone.

The goal is to use the self-assembly of the nucleic acid structures to template the assembly of the nanoparticles hosted on them, controlling their position and in some cases orientation. Many of these schemes use a covalent attachment scheme, using oligonucleotides with amide or thiol functional groups as a chemical handle to bind the heteroelements.

This covalent binding scheme has been used to arrange gold nanoparticles on a DX-based array, and to arrange streptavidin protein molecules into specific patterns on a DX array. A non-covalent hosting scheme using Dervan polyamides on a DX array was used to arrange streptavidin proteins in a specific pattern on a DX array.

Carbon nanotubes have been hosted on DNA arrays in a pattern allowing the assembly to act as a molecular electronic device, a carbon nanotube field-effect transistor.

In addition, there are nucleic acid metallization methods, in which the nucleic acid is replaced by a metal which assumes the general shape of the original nucleic acid structure, and schemes for using nucleic acid nanostructures as lithography masks, transferring their pattern into a solid surface.

Dynamic DNA Nanotechnology

Dynamic DNA nanotechnology focuses on creating nucleic acid systems with designed dynamic functionalities related to their overall structures, such as computation and mechanical motion.

There is some overlap between structural and dynamic DNA nanotechnology, as structures can be formed through annealing and then reconfigured dynamically, or can be made to form dynamically in the first place.

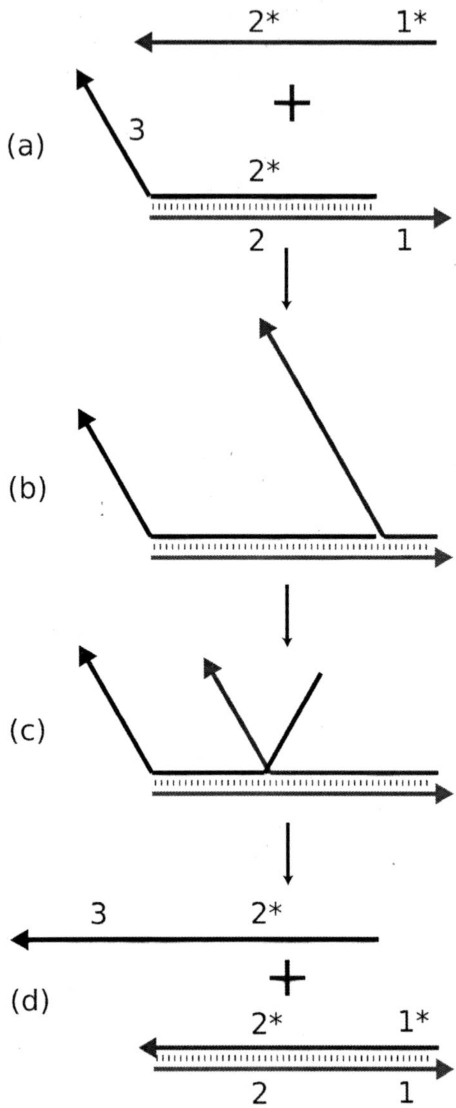

Figure: Dynamic DNA nanotechnology often makes use of toehold-mediated strand displacement reactions. In this example, the red strand binds to the single stranded toehold region on the green strand (region 1), and then in a branch migration process across region 2, the blue strand is displaced and freed from the complex. Reactions like these are used to dynamically reconfigure or assemble nucleic acid nanostructures. In addition, the red and blue strands can be used as signals in a molecular logic gate.

Nanomechanical Devices

DNA complexes have been made that change their conformation upon some stimulus, making them one form of nanorobotics. These structures are initially formed in the same way as the static structures made in structural DNA nanotechnology, but are designed so that dynamic reconfiguration is possible after the initial assembly. The earliest such device made use of the transition between the B-DNA and Z-DNA forms to respond to a change in buffer conditions by undergoing a twisting motion. This reliance on buffer conditions, however, caused all devices to change state at the same time. Subsequent systems could change states based upon the presence of control strands, allowing multiple devices to be independently operated in solution. Some examples of such systems are a "molecular tweezers" design that has an open and a closed state, a device that could switch from a paranemic-crossover (PX) conformation to a double-junction (JX2) conformation, undergoing rotational motion in the process, and a two-dimensional array that could dynamically expand and contract in response to control strands. Structures have also been made that dynamically open or close, potentially acting as a molecular cage to release or reveal a functional cargo upon opening.

DNA walkers are a class of nucleic acid nanomachines that exhibit directional motion along a linear track. A large number of schemes have been demonstrated. One strategy is to control the motion of the walker along the track using control strands that need to be manually added in sequence. Another approach is to make use of restriction enzymes or deoxyribozymes to cleave the strands and cause the walker to move forward, which has the advantage of running autonomously. A later system could walk upon a two-dimensional surface rather than a linear track, and demonstrated the ability to selectively pick up and move molecular cargo. Additionally, a linear walker has been demonstrated that performs DNA-templated synthesis as the walker advances along the track, allowing autonomous multistep chemical synthesis directed by the walker.

Strand Displacement Cascades

Cascades of strand displacement reactions can be used for either computational or structural purposes. An individual strand displacement reaction involves revealing a new sequence in response to the presence of some initiator strand. Many such reactions can be linked into a cascade where the newly revealed output sequence of one reaction can initiate another strand displacement reaction elsewhere. This in turn allows for the construction of chemical reaction networks with many components, exhibiting complex computational and information processing abilities. These cascades are made energetically favourable through the formation of new base pairs, and the entropy gain from disassembly reactions. Strand displacement cascades allow for isothermal operation of the assembly or computational process, as opposed to traditional nucleic acid assembly's requirement for a thermal annealing step, where the temperature is raised and then slowly lowered to ensure proper formation of the desired structure. They can also support catalytic functionality of the initiator species, where less than one equivalent of the initiator can cause the reaction to go to completion.

Strand displacement complexes can be used to make molecular logic gates capable of complex computation. Unlike traditional electronic computers, which use electric current as inputs and outputs, molecular computers use the concentrations of specific chemical species as signals. In the case of nucleic acid strand displacement circuits, the signal is the presence of nucleic acid strands that are released or consumed by binding and unbinding events to other strands in displacement complexes. This approach has been used to make logic gates such as AND, OR, and NOT gates. More recently, a four-bit circuit was demonstrated that can compute the square root of the integers 0–15, using a system of gates containing 130 DNA strands.

Another use of strand displacement cascades is to make dynamically assembled structures. These use a hairpin structure for the reactants, so that when the input strand binds, the newly revealed sequence is on the same molecule rather than disassembling. This allows new opened hairpins to be added to a growing complex. This approach has been used to make simple structures such as three- and four-arm junctions and dendrimers.

Applications

DNA nanotechnology provides one of the few ways to form designed, complex structures with precise control over nanoscale features. The field is beginning to see application to solve basic science problems in structural biology and biophysics. The earliest such application envisaged for the field, and one still in development, is in crystallography, where molecules that are difficult to crystallize in isolation could be arranged within a three-dimensional nucleic acid lattice, allowing determination of their structure. Another application is the use of DNA origami rods to replace liquid crystals in residual dipolar coupling experiments in protein NMR spectroscopy; using DNA origami is advantageous because, unlike liquid crystals, they are tolerant of the detergents needed to suspend membrane proteins in solution. DNA walkers have been used as nanoscale assembly lines to move nanoparticles and direct chemical synthesis. Furthermore, DNA origami structures have aided in the biophysical studies of enzyme function and protein folding.

DNA nanotechnology is moving towards potential real-world applications. The ability of nucleic acid arrays to arrange other molecules indicates its potential applications in molecular scale electronics. The assembly of a nucleic acid structure could be used to template the assembly of a molecular electronic elements such as molecular wires, providing a method for nanometre-scale control of the placement and overall architecture of the device analogous to a molecular breadboard. DNA nanotechnology has been compared to the concept of programmable matter because of the coupling of computation to its material properties.

In a study conducted by a group of scientists from iNANO centre and CDNA Centre in Aarhus university (Aarhus), researchers were able to construct a small multi-switchable 3D DNA Box Origami. The proposed nanoparticle was characterized by AFM, TEM and FRET. The constructed box was shown to have a unique reclosing mechanism, which enabled it to repeatedly open and close in response to a unique set of DNA or RNA keys. The authors proposed that this "DNA device can potentially be used for a broad range of applications such as controlling the function of single molecules, controlled drug delivery, and molecular computing.".

There are potential applications for DNA nanotechnology in nanomedicine, making use of its ability to perform computation in a biocompatible format to make "smart drugs" for targeted drug delivery. One such system being investigated uses a hollow DNA box containing proteins that induce apoptosis, or cell death, that will only open when in proximity to a cancer cell. There has additionally been interest in expressing these artificial structures in engineered living bacterial cells, most likely using the transcribed RNA for the assembly, although it is unknown whether these complex structures are able to efficiently fold or assemble in the cell's cytoplasm.

If successful, this could enable directed evolution of nucleic acid nanostructures. Scientists at Oxford University reported the self-assembly of four short strands of synthetic DNA into a cage which is capable of entering cells and surviving for at least 48 hours. The fluorescently labelled DNA tetrahedrons were found to remain intact in the laboratory cultured human kidney cells despite the attack by cellular enzymes after two days. This experiment showed the potential of drug delivery inside the living cells using the DNA 'cage'. A DNA tetrahedron was used to deliver RNA Interference (RNAi) in a mouse model, reported a team of researchers in MIT. Delivery of the interfering RNA for treatment has showed some success using polymer or lipid, but there are limitations of safety and imprecise targeting, in addition to short shelf life in the blood stream. The DNA nanostructure created by the team consists of six strands of DNA to form a tetrahedron, with a single strand of RNA affixed to each of the six edges. The tetrahedron is further equipped with targeting protein, three folate molecules, which lead the DNA nanoparticles to the abundant folate receptors found on some tumors. The result showed that the gene expression targeted by the RNAi, luciferase, dropped by more than half. This study shows promise in using DNA nanotechnology as an effective tool to deliver treatment using the emerging RNA Interference technology.

Materials and Methods

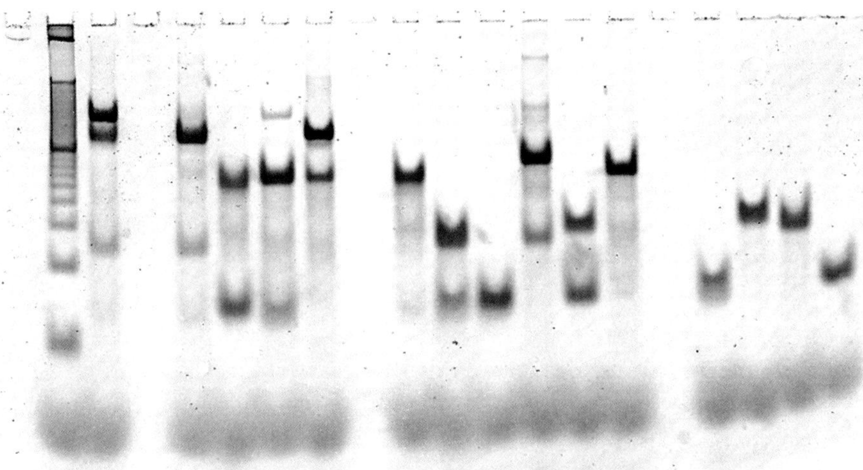

***Figure:** Gel electrophoresis methods, such as this formation assay on a DX complex, are used to ascertain whether the desired structures are forming properly. Each vertical lane contains a series of bands, where each band is characteristic of a particular reaction intermediate.*

The sequences of the DNA strands making up a target structure are designed computationally, using molecular modelling and thermodynamic modelling software. The nucleic acids themselves are then synthesized using standard oligonucleotide synthesis methods, usually automated in an oligonucleotide synthesizer, and strands of custom sequences are commercially available. Strands can be purified by denaturing gel electrophoresis if needed, and precise concentrations determined via any of several nucleic acid quantitation methods using ultraviolet absorbance spectroscopy.

The fully formed target structures can be verified using native gel electrophoresis, which gives size and shape information for the nucleic acid complexes. An electrophoretic mobility shift assay can assess whether a structure incorporates all desired strands. Fluorescent labelling and Förster resonance energy transfer (FRET) are sometimes used to characterize the structure of the complexes. Nucleic acid structures can be directly imaged by atomic force microscopy, which is well suited to extended two-dimensional structures, but less useful for discrete three-dimensional structures because of the microscope tip's interaction with the fragile nucleic acid structure; transmission electron microscopy and cryo-electron microscopy are often used in this case. Extended three-dimensional lattices are analyzed by X-ray crystallography.

Nanotechnology of Diamond

The nanodiamond, also sometimes called a detonation nanodiamond (DND) or ultradispersed diamond (UDD), is produced as the result of a carefully controlled explosion. Nanodiamonds are usually extracted from the explosion's resulting soot through the use of pressure, heat, and acid.

Alternately, nanodiamonds can also be produced using graphite-based processes. These tiny diamonds are used in industrial applications involving a wide range of products. Beginning in the late 20th century, research in medical applications revealed that nanodiamonds are effective agents for both delivering and monitoring chemotherapy drugs.

In order to manufacture nanodiamonds, carbon is subjected to the detonation of a TNT and RDX mixture, two powerful explosives, resulting in the production of a sooty material containing nanodiamonds of about 5 nanometres (nm). The vessel containing the explosion is rapidly cooled using compressed carbon dioxide, water, or air, as faster cooling produces a larger diamond yield.

The diamond-laden soot is then placed in an autoclave and boiled in acid under high pressure to remove almost all of the impurities such as the contaminating metals from the explosion vessel as well as lower-grade carbon incapable of producing diamonds. Nanodiamonds can also be synthesized using graphite in suspension that is either subjected to ultrasonic cavitation or a pulsating laser beam.

Nanodiamonds possess a very large surface area relative to their size so that hydrocarbon and water molecules attach readily to them. The nanodiamond particles also tend to cluster and adhere to each other strongly. These characteristics make handling them challenging while also giving them many potential industrial and medical applications. Industrially, the nanodiamond is used in dry lubricants, reinforcement for plastics and rubber, and polishing, as well as lapping processes and as an additive to motor oils.

A group of researchers at Northwestern University in Evanston, Illinois, have demonstrated that nanodiamonds offer great medical potential when used in conjunction with anti-cancer chemotherapy drugs. Due to their purity, the diamonds are not attacked by the immune system while binding strongly to molecules containing medicines, thus overcoming the tendency of tumors to resist medicines. Due to its tiny size, the nanodiamond is easily flushed by the urinary tract following chemotherapy treatments.

Tests on laboratory mice with drug-resistant cancers have demonstrated that drugs delivered with nanodiamonds remain in the bloodstream more than ten times longer, producing substantially enhanced tumor shrinkage.

By adding contrast agents to nanodiamonds and using magnetic resonance imaging (MRI), the Northwestern researchers have also shown that they can more precisely track the location and effects of chemotherapy drugs.

Nano Diamonds

The Technology - The production technology employs "detonation synthesis" to create diamond particles with average sizes of 4-6 nm (nanometres) each. A supplementary proprietary refinement process further modifies the results of the detonation synthesis stage to extract various grades of these "nanodiamonds" (also referred to as "ultra dispersed diamonds – UDD) that meet technical requirements for various applications.

The basic idea of creating the technology for production of nanodiamonds was based on fundamental achievements of Soviet military science from experiments with explosive substances, for example, solid rocket fuels. Based on these experiments Soviet military science learned to make ultra dispersed diamonds. The size of the crystals of such diamonds is one thousands times smaller than other synthetic diamonds. This is reputedly a new class of materials with unique and commercially valuable characteristics.

The detonation synthesis technology employs charges of explosive substances which are detonated in high strength round hermetically sealed chambers. Diamond particles are formed from the free carbon of the molecules of the explosives at temperatures of approximately 3,500 degrees C. and at pressures of approximately 200,000 atmospheres.

The detonation products (diamond blend – DB) contain, on average, 40 – 50% diamonds and are subjected to a subsequent chemical purification process.

Physical

Phisical	
Mean Crystal Size	4 – 8 nanometers (0.004 - 0.008 microns)
Aggregate size	20 – 50 nanometer (0.02 – 0.05 microns)
Specific Surface	300 m2/g
Density	3.1 – 3.2 g/cm3
Bulk Weight	0.4 – 0.6 g/cm3
Lattice Constant	0.3573 +0.0005 nanometers
Air Oxidation Starting Temperature	450°C
Vacuum Graphitization Starting Temperature	1100°C
Chemical composition:	Carbon (88%-92%) Nitrogen (1%-3%) Hydrogen (1%-2%) Oxygen (0%-10%)
Physical composition:	Regular UDD (80%-100% Hexagonal UDD (1%-20%)

UDD of Different Grades are Used.

- in electrochemical plating with base metals – chromium, nickel, zinc, copper, tin, etc.;
- in electrochemical plating with noble metals – gold, platinum, silver, etc.;
- in anodic oxidation of aluminum and aluminum alloys;
- for polishing with micro-abrasive and finishing compositions;
- in lubricants – oils and greases, cooling and lubricating liquids (antifrictional wear-protector lubricating compositions);
- in polymers (films, rubbers, plastics) and membranes;
- in abrasive tools;
- in magnetic recording and microelectronic systems;
- in electronics (diamond substrate – high dispersive seeds in production of diamond films and body-centred structures using UDD laser evaporation and epitaxy techniques);
- in medicine;
- for analysis purposes (high resolution adsorbents), etc.

Electrochemical Plating with Base Metals

Fields of application for chromium, nickel, zinc, copper, tin-based electroplated coatings:

- parts, tools, automobile bodies;
- moulds, pressing tooling;
- medical instruments;
- electrical engineering, radio and electronic industry;
- magnetic data media.

Metal-UDD coatings provide high degree of protection against corrosion and mechanical wear and attractive appearance of the products.

Coatings have uniform crystal structure, stable electrical and physical properties and are not harmful for a user.

Metal-UDD coatings may be produced on a standard equipment after its minimal modification.

Advantages of Metal-UDD Coatings:

- Improved quality and competitiveness of the products, viz.:
  - improved microhardness
    (Cr-UDD – to 1400 kg/mm^2, Ni-UDD – to 850 kg/mm2 , Sn-UDD – to 20 kg/mm2)
  - improved wear resistance;
  - increased adhesion;
  - lower friction coefficient;
  - coatings having no pores or fracturing at lower thickness.
- 2 to 15 times longer service time of the parts and tools.
- Saving of materials, power and labor.
- Lower thickness of coatings and higher rate of deposition result in increased productivity of plating lines (by 20 to 50%).
- Improvement of environment protection characteristics.

Electrochemical Plating with Noble Metals.

Fields of application for gold and silver-based electroplated coatings:

- watch and jewelry production;
- medicine;
- electrical engineering, radio and electronic industry.

UDD-containing coatings improve appearance of the products and ensure high degree of protection against corrosion and mechanical wear. Coatings have uniform microcrystal structure, stable electrical and physical properties and are not harmful for a user. Metal-UDD coatings may be produced on a standard equipment after its minimal modification.

Advantages of Metal-UDD Coatings:

- Improved quality and competitiveness of the products, viz.:
  - improved abrasive resistance
  - improved microhardness

    (Ag-UDD – to 180 kg/mm2, Au-UDD – to 200 kg/mm2)
  - increased adhesion;
- Increased service time of the products.
- Lower thickness of coatings and higher rate of deposition result in increased productivity of plating lines (by 20 to 50%).
- Coatings of lower thickness that guarantee the same features and require similar power and labor resources enable saving of noble metals.
- Improvement of environment protection characteristics.

Anodic Oxidation of Aluminum and Aluminum Alloys

Fields of application for anodic oxide UDD containing coatings of aluminum and aluminum alloys:

- electrical engineering;
- radio electronic industry;
- electronic industry;
- machine-building industry.

Anodic oxide UDD containing coatings provide high hardness and wear resistance of the products, improve their electrical and heat insulation properties, corrosion resistance and appearance. Filling of such coatings is 3 times higher than that of standard coatings and they are not harmful for a user. Process of the anodic oxidation with UDD is performed on a standard equipment in different electrolytes.

Advantages of Anodic Oxide UDD Containing Coatings:

- Improved quality and competitiveness of the products, viz.:
  - improved microhardness;
  - improved wear resistance;
  - high filling and low porosity;

- increased adhesion;
- lower friction coefficient;
- improved electrical and heat insulation properties.

- Increased service time of the products.
- Saving of materials, power and labor.

Lower thickness of coatings and higher rate of deposition result in increased productivity of plating lines (by 20 to 50%).

Abrasive Tools.

There are two directions for application of the UDD for abrasive tools:

1. Compact (sintered) polycrystal product – as polishing, finishing and cutting tools;
2. Polymer binder strengthener for elastic abrasive compositions and tools.

Fields of application for UDD:

- abrasive industry;
- mining industry;
- metal working;
- machine building;
- electronics;
- medicine;
- jewelry production.

1. Quality Features of Compact Cakes:

Pure UDD:

- microhardness of 3.0 – 3.5 thousands kg/mm2;
- R = 1012 – 1014 Ohm.cm;?g = 100 – 150 W/m k, e = 7 - 9
- density of defect-free fragments of 3.35 – 3.50 g/cm3

Oxide-UDD compositions with vitrifying oxides (UDD ceramics):

- microhardness of 3.0 – 4.0 thousands kg/mm2 (UDD- MgO);
- R = 1012 – 1014 Ohm.cm; g = 120 – 160 W/m k, e = 8 -10

These compacts are easily cut with ultrasonic and laser equipment, ground and polished. The compositions may be used in machining of materials of medium hardness, as well as in electronics.

Metallized UDD with transition metal clusters:

- microhardness of 6.0 – 7.0 thousands kg/mm^2.

A very good metal for tools.

UDD (0.5 – 1.5%) used to increase strength parameters of the polymer matrix in the polishing and grinding compositions also bring about increase both in strength properties (1.3 – 1.5 times), and elasticity properties (1.8 – 2.0 times) of the abrasive tools.

Ultra Dispersive Diamonds (UDD, Nanodiamonds) for Magnetic Recording Systems

Ways of using UDD:

- as antifrictional additive and physical modifier in ferromagnetic layer of the magnetic tapes and discs;
- as an additive in electrochemical composite films to improve characteristics of the magnetic recording systems.

1. Addition of UDD to the ferromagnetic layer causes drastic decrease of the domain extent (ferromagnetic grains), resulting in the increased recording density. When UDD are added to the varnishes for special tapes used for the magnetic head cleaning, the magnetic heads are polished and their wear resistance substantially increases.

Advantages of UDD containing soft magnetic data media:

- decrease in abrasive wear of the magnetic medium layer;
- optimal operation mode is provided for the magnetic heads and playback devices;
- decreased friction in the magnetic medium and increased stability of its movement

2. a) Magnetoosoft amorphous CoP-UDD coatings

Advantages compared to pure CoP coatings:

- 1.3 times increase in microhardness;
- 4.5 times increase in wear resistance;
- 1.4 times decrease in friction coefficient;
- magnetic characteristics remain as before;
- increased service time (by 2 times) of the head cores without changing of their efficiency

b) Magnetohard polycrystalline CoP-UDD coatings:

Advantages compared to pure CoP coatings:

- 1.2 times increase in microhardness;
- 1.6 times decrease in corrosion currents;

- magnetic characteristics remain as before;
- increased service time of the magnetic recording media.

Medicine

Significant number of unpaired electrons on the UDD surface makes each diamond crystal a powerful multicharge donor radical capable of "quenching" intensive radical processes which are present in actually all heavy human diseases. Individual UDD crystal has a capability to penetrate the walls of a live cell and to interact there with active radicals, neutralizing pathogenic bacteria and viruses.

UDD possess extremely high activity respect to bacteria and viruses, absorbing and thus killing them. UDD are a new generation of antiseptics of specific action.

UDD are super-active sorbents, immobilizers of biologically active substances.

UDD are capable of enhancing the effect of drugs. Using of UDD normalize blood pressure.

UDD are also effective in diseases of the gastrointestinal tract.

UDD are an effective aid to heal post-burn conditions, various skin diseases, intoxications.

Application of the UDD aqueous and oil suspensions for cancer patients contrubutes to:

- enhancement of the medication efficiency;
- relief of the pains;
- improvement of the CBC parameters;
- stimulation of tonus;
- discharge of toxins;
- prolonging patient's life.

UDD are not mutagenic or cancerogenic.

For one course of treatment 0.05 g of UDD is required.

Polishing with micro-abrasive and finishing compositions.

Application of the UDD polishing systems:

- high-precision finishing of materials used in radio engineering, electronics, optics, medicine, machine-building.

The compositions enable to obtain ideal mirror surface of solids, without any defects or dislocations, with required planeness, plane-parallelness, of any geometrical shape, with roughness relief height Rz = 28 nm (a = 0.5 to 2 nm).

Advantages of UDD containing polishing systems:

- ultra-small UDD size ensures minimal surface roughness and colloidal stability of the polishing systems
- chemical stability of UDD allows to use chemically active additives for the polishing systems and enables their regeneration;
- small material takeover from the polishing surface;

Ion-exchange and sorption activity of the UDD provides immobilization of ion and molecular products of

takeover from the UDD surface;

- fractional structure of the UDD aggregates provides a regulated structure formation in the suspension polishing compositions and guarantees absence of any strained or broken surface layers of the material being treated;
- UDD polishing systems are not toxic;
- UDD containing polishing systems mean improved quality and competitiveness of the polishing compositions and tools, technological effectiveness of polishing hard-in-treatment materials.

UDD (Nanodiamonds) in Polymers

Application of the UDD-based polymers:

- aircraft, automobile, tractor ans ship-building;
- medicine;
- chemical and oil chemistry industry;
- production of seals, stop fittings of any kind;
- protector and antifrictional film coatings.

Polymer compositions are characterized by enhanced complex of elasticity-strength properties (1.5 – 2.0 times). UDD containing fluoroelastomers have increased abrasive wear resistance (2 – 4 times) and films produced of those have dry friction coefficient less than 0.01. UDD are an effective stabilizing agent for thermal ageing og the polymers. UDD-based epoxy adhesives are characterized by high adhesive and cohesive properties.

UDD are used in the technologies of solution and melting polymerization, chemical solidification, cathode ray, gas flame and electrostatic atomization. UDD-containing composites are produced on standard technological equipment.

Advantages of the UDD-based polymers compositions:

- Improved quality and competitiveness of the products, viz.:
  - improved strength wear resistance, ageing and abrasive resistance;
  - lower friction coefficient for polyfluoroelastomeres and perfluoropolymers; high friction coefficient for polyizoprenes;
  - adaptability of materials and coatings to specific technologies and products;
  - possibility to replace expensive polymer raw materials with cheap ones ;
  - saving of expensive and/or deficit components and materials.

Lubricants – oils and greases, cooling and lubricating liquids.

Application of the UDD additives in lubricants, greases and cooling and lubricating liquids:

- machine-building;
- metal cutting and treatment;
- motor building;
- ship building;
- aircraft building;
- transport facilities, etc.

Advantages of the UDD additives for lubricating compositions:

- Improved quality and competitiveness of the products, viz.:
  - increased operation time of transport means and equipment;
  - saving of fuel and lubricants;
  - a technology of choice for expensive lubricants and specialised greases;
- Lowered frictional torque (20 to 40%).
- Lowered wear of frictional surfaces (30 - 40% and more).
- Quick run-in of friction pairs.

Detonation Nanodiamond

Detonation nanodiamond (DND), often also called ultradispersed diamond (UDD), is diamond that originates from a detonation. When an oxygen-deficient explosive mixture of TNT/RDX is detonated in a closed chamber, diamond particles with a diameter of ca. 5 nm are formed at the front of detonation wave in time of several microseconds.

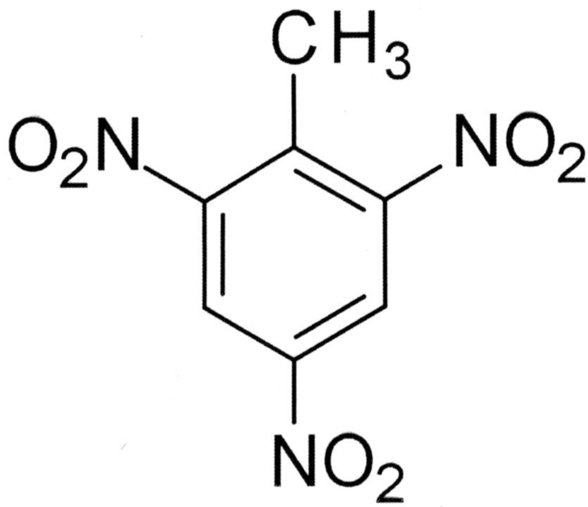

Figure: *trinitrotoluene (TNT) structure*

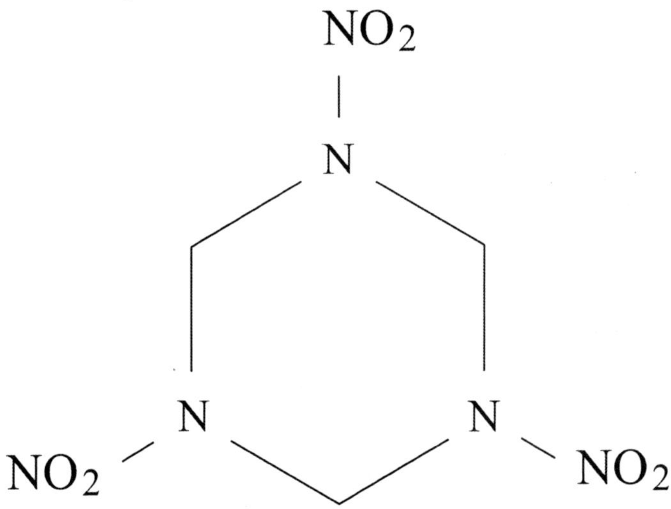

Figure: *hexogen (RDX) stru*

Detonation nanodiamonds were first synthesized in 1962 by a group of Soviet scientists at VNIITF led by Yevgeny Zababakhin, including K. V. Volkov, Vyacheslav Danilenko, and V. I. Elina.

Properties

The diamond yield after detonation crucially depends on the synthesis condition and especially on the heat capacity of the cooling medium in the detonation chamber (water, air, CO_2, etc.). The higher the cooling capacity, the larger the diamond yield, which can reach 90%. After the synthesis, diamond is extracted from the soot using high-temperature high-pressure (autoclave) boiling in acid for a long period (ca. 1–2 days). The boiling removes most of the metal contamination, originating from the chamber materials, and non-diamond carbon. Various measurements, including X-ray diffraction and high-resolution transmission electron microscopy revealed that the size of the diamond grains in the soot is distributed around 5 nm.

The grains are unstable with respect to aggregation and spontaneously form micrometre-sized clusters. The adhesion is strong and contacts between a few nano-grains can hold a micrometre-sized cluster attached to a substrate.

Nanosized diamond has extremely large relative surface area. As a result, its surface spontaneously attaches water and hydrocarbon molecules from the ambient atmosphere. However, clean nanodiamond surface can be obtained with appropriate handling.

The detonation nanodiamond grains mostly have diamond cubic lattice and are structurally imperfect. The major defects are multiple twins, as suggested by High-resolution transmission electron microscopy. Despite the carbon source for the diamond synthesis - TNT/RDX explosive mixture - being rich in nitrogen, concentration of paramagnetic nitrogen inside diamond grains is below 1 part per million (ppm). Paramagnetic nitrogen (neutral nitrogen atoms substituting carbon in the diamond lattice) is the major form of nitrogen in diamond, and thus the nitrogen content in DND is probably very low.

Alternative Synthesis Methods

Diamond nanocrystals can also be synthesized from a suspension of graphite in organic liquid at atmospheric pressure and room temperature using ultrasonic cavitation. The yield is approximately 10%. The cost of nanodiamonds produced by this method are estimated to be competitive with the HPHT process.

An alternative synthesis technique is irradiation of graphite by high-energy laser pulses. The structure and particle size of the obtained diamond is rather similar to that obtained in explosion. In particular, many particles exhibit multiple twinning.

Recently Mohan Sankaran and Ajay Kumar from Case Western Reserve University achieve the fabrication of nanodiamonds 2–5nm in size at near-ambient conditions by a microplasma process. Instead of having to use tons of crushing force and volcanic heat to forge diamonds, researchers at Case Western Reserve University have developed a way to cheaply make nanodiamonds on a lab bench at atmospheric pressure and near room temperature.

The nanodiamonds are formed directly from a gas and require no surface to grow on.

Applications

Commercial products based on nanodiamonds are available for the following applications:

1. Lapping and polishing (e.g. Sufipol);
2. Additives to engine oils (e.g. ADDO);
3. Dry lubricants for metal industry (Drawing of W-, Mo-, V-, Rh-wires);
4. Reinforcing fillers for plastic and rubber;
5. Additives to electroplating electrolyte (e.g. DiamoSilb, DiamoChrom).

Use in Medicine

Prof. Dean Ho, Dr. Houjin Huang, and Dr. Erik Pierstorff at Northwestern University, in collaboration with Dr. Eiji Osawa at The NanoCarbon Research Institute in Japan, have demonstrated that nanomaterials can shuttle chemotherapy drugs to cells without producing the negative effects of today's delivery agents.

Clusters of the nanodiamonds surround the drugs ensuring that they remain separated from healthy cells, preventing unnecessary damage; upon reaching the intended targets, the drugs are released into the cancer cells. The leftover diamonds, hundreds of thousands of which could fit into the eye of a needle, do not induce inflammation in cells once they have done their job.

Nanotechnology in Space

Nanotechnology will play an important role in future space missions. Nanosensors, dramatically improved high-performance materials, or highly efficient propulsion systems are but a few examples.

Propulsion Systems

Most of today's rocket engines rely on chemical propulsion. All current spacecraft use some form of chemical rocket for launch and most use them for attitude control as well (the control of the angular position and rotation of the spacecraft, either relative to the object that it is orbiting, or relative to the celestial sphere). Real rocket scientists though are actively researching new forms of space propulsion systems.

One heavily researched area is electric propulsion (EP) that includes field emission electric propulsion (FEEP), colloid thrusters and other versions of field emission thrusters (FETs). EP systems significantly reduce the required propellant mass compared to conventional chemical rockets, allowing to increase the payload capacity or decrease the launch mass. EP has been successfully demonstrated as primary propulsion systems for NASA's Deep Space 1, Japan's HAYABUSA, and ESA's SMART-1 missions.

A nanotechnology EP concept proposes to utilise electrostatically charged and accelerated nanoparticles as propellant. Millions of micron-sized nanoparticle thrusters would fit on one square centimetre, allowing the fabrication of highly scalable thruster arrays.

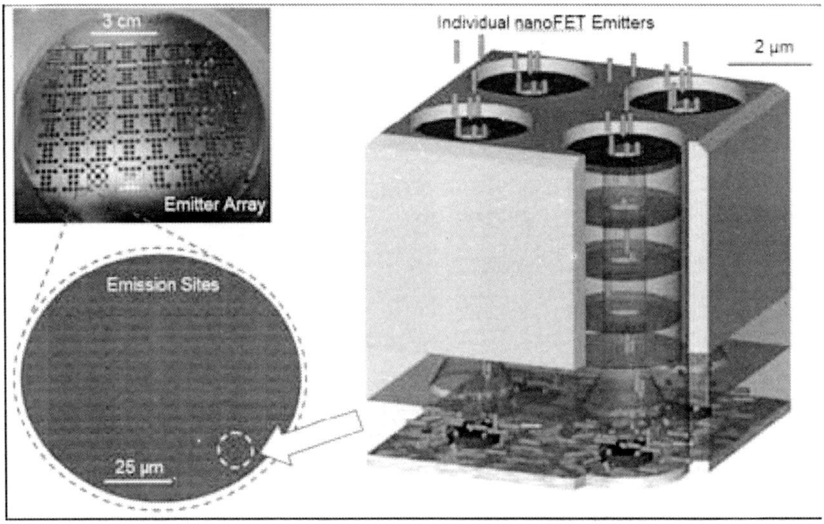

Figure: *nanoFET characteristic size scales (Image: University of Michigan Department of Aerospace Engineering)*

Pretty far out are proposals that the manipulation of Casimir forces could lead to a propulsion system for interstellar spaceships. The basic idea is that if one could exploit the fact that vacuum is an energy reservoir, thanks to zero-point energy, future space travelers would have access to a limitless energy source. The only thing they need, of course, is some kind of propulsion system that harvests the required energy from the vacuum. That this is not totally crazy was demonstrated in a 1984 paper. Serious research efforts are being made in various laboratories to harness the Casimir and related effects for vacuum energy conversion (read more: "nanotechnology, the mysterious Casimir Force, and interstellar spaceships").

Radiation Shielding

Radiation shielding is an area where nanotechnology could make a major contribution to human space flight. NASA says that the risks of exposure to space radiation are the most significant factor limiting humans' ability to participate in long-duration space missions. A lot of research therefore focuses on developing countermeasures to protect astronauts from those risks. To meet the needs for radiation protection as well as other requirements such as low weight and structural stability, spacecraft designers are looking for materials that help them develop multifunctional spacecraft hulls. Advanced nanomaterials such as the newly developed, isotopically enriched boron nanotubes could pave the path to future spacecraft with nanosensor-integrated hulls that provide effective radiation shielding as well as energy storage.

Another area of required radiation shielding is the protection of onboard electronics. It has been reported previously that electronic devices became more radiation tolerant when their dimensions are reduced. For example, multi-quantum well or quantum dot devices can be tens or hundreds times more radiation tolerant than conventional bulk devices. It even was shown that quantum dot/CNT-based photovoltaic devices were five orders of magnitude more resistant than conventional bulk solar cells.

Recently, a few studies on radiation effects of high energetic particles such as proton, electron, and heavy ions on nanomaterials like carbon nanotubes and nanowires have focused on the changed structural properties of the nanomaterials after irradiation.

Anti-Satellite Weapon Countermeasure

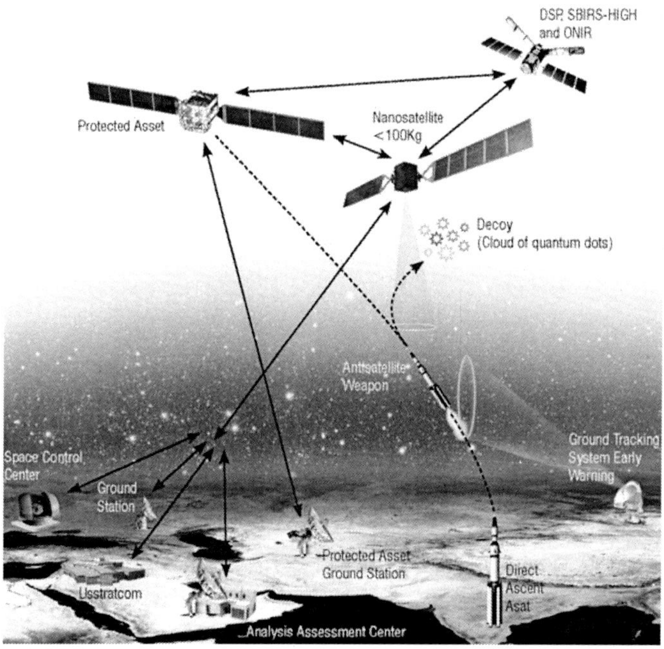

Figure: *Schematic of a countermeasure system: Anti-satellite weapon seeks the decoy cloud of quantum dots instead of the target satellite.*

In January 2007, China successfully tested an Anti-satellite (ASAT) missile system by destroying their own defunct LEO satellite, which generated huge amounts of space debris. This ASAT test raised worldwide concerns about the vulnerability of satellites and other space assets and possibility of triggering an arms race in space.

In order to meet emerging challenges posed by such ASAT missile systems, military strategists and researchers arc developing novel technologies to protect their space assets. In view of this, Raytheon Company has developed a counter measure system using quantum dots to protect space assets such as satellites from missile attacks. They have developed a decoy consisting of quantum dots of different sizes and shapes that are engineered to emit radiation having a radiation profile similar to that of the asset.

Space Elevator

Tie a rock to the end of a piece of ribbon, then spin it over your head. It will be pulled taut as the rock circles about. Now, imagine a ribbon 62,000 miles long, anchored near the equator with a weight on the other end.

The centrifugal force of the earth's rotation will make it behave the same way. You'll end up with not only the world's biggest nunchuck, but also a kind of elevator to outer space. A space elevator based on carbon nanotube cables is one of those ideas from 1950s-style futurism that are so whacky they might just work.

Protecting Satellites from Energy Weapons

A recent research paper published at the Centre for Strategy and Technology, at the Air Force's Air War College, discusses how nanotechnology can be used to improve the design of satellites to mitigate the threats posed by ground-based directed energy weapons and high-powered microwaves. The paper states that several nations, including the U.S., Russia and China, already have either built or are developing the technology to construct ground-based directed energy weapons.

Space Instrumentation

Black is black, right? Not so, according to a team of NASA engineers now developing a blacker-than pitch nanomaterial that will help scientists gather hard-to-obtain scientific measurements or observe currently unseen astronomical objects, like Earth-sized planets in orbit around other stars. The nanomaterial being developed by a team of 10 technologists at the NASA Goddard Space Flight Centre in Greenbelt,

Md., is a thin coating of multi-walled carbon nanotubes. While carbon nanotubes would find use in the Space Elevator thanks to their extraordinary strength, in this application, NASA is interested in using the technology to help suppress errant light that has a funny way of ricocheting off instrument components and contaminating measurements.

MIT hosts the Space Nanotechnology Laboratory whose primary mission is to develop nano-fabrication, advanced lithography and precision engineering technology for building high performance space instrumentation, including x-ray telescopes and high resolution x-ray spectrometres, magnetospheric imagers and solar physics instrumentation.

6

Thinfilm Optics

Nanophotonics or Nano-optics is the study of the behaviour of light on the nanometre scale. It is considered as a branch of optical engineering which deals with optics, or the interaction of light with particles or substances, at deeply subwavelength length scales. Technologies in the realm of nano-optics include near-field scanning optical microscopy (NSOM), photoassisted scanning tunnelling microscopy, and surface plasmon optics. Traditional microscopy makes use of diffractive elements to focus light tightly in order to increase resolution. But because of the diffraction limit (also known as the Rayleigh Criterion), propagating light may be focused to a spot with a minimum diameter of roughly half the wavelength of the light. Thus, even with diffraction-limited confocal microscopy, the maximum resolution obtainable is on the order of a couple of hundred nanometres. The scientific and industrial communities are becoming more interested in the characterization of materials and phenomena on the scale of a few nanometres, so alternative techniques must be utilised. Scanning Probe Microscopy (SPM) makes use of a "probe", (usually either a tiny aperture or super-sharp tip), which either locally excites a sample or transmits local information from a sample to be collected and analyzed. The ability to fabricate devices in nanoscale that has been developed recently provided the catalyst for this area of study.

The study of nanophotonics involves two broad themes 1) studying the novel properties of light at the nanometre scale 2) enabling highly power efficient devices for engineering applications.

The study has the potential to revolutionize the telecommunications industry by providing low power, high speed, interference-free devices such as electrooptic and all-optical switches on a chip.

In detail

The term typically refers to phenomena of ultraviolet, visible and near IR light, with a wavelength of approximately 300 to 1200 nanometres.

The interaction of light with these nanoscale features leads to confinement of the electromagnetic field to the surface or tip of the nanostructure resulting in a region referred to as the optical near field. This effect is to some extent analogous to a lightning rod, where the field concentrates at the tip. In this region, the field may need to adjust to the topography of the nanostructure. This means that the electromagnetic field will be dependent on the size and shape of the nanostructure that the light is interacting with.

This optical near field can also be described as a surface bound optical oscillation which can vary on length scale of tens or hundreds of nanometres – a length scale smaller than the wavelength of the incoming light. This can provide higher spatial resolution beyond the limitations imposed by the law of diffraction in conventional far-field microscopy. The technique derived from this effect is known as near-field microscopy, and opens up many new possibilities for imaging and spectroscopy on the nanoscale. A novel embodiment which has picometre resolution in the vertical plane above the waveguide surface is dual polarisation interferometry.

Novel optical properties of materials can result from their extremely small size. A typical example of this type of effect is the colour change associated with colloidal gold. In contrast to bulk gold, known for its yellow colour, gold particles of 10 to 100 nm in size exhibit a rich red colour. The critical size where these and related effects take place are correlated with the mean free path of the conduction electrons of the metal.

In addition to these extrinsic size effects that determine a material's optical response to incoming light, the intrinsic properties of the material can change. These size effects occur as particles become even smaller. At this stage some of the intrinsic electronic properties of the medium itself change. One example of this phenomenon is in semiconductor nanostructures where the extremely small particle size confines the quantum mechanical wavefunction, leading to discrete optical transitions, e.g., fluorescence colours that depend on the size of the particle. The changing bandgap of the semiconductor is the reason for this colour change. This effect, however, since not directly correlated with optical wavelength, is not unanimously included when referring to nano-optics.

In March 2010, S. Assefa *et al.* of IBM reported invention of ultra-fast and noise-free nanophotonic avalanche photodetectors which are poised to bring about the exaflop light circuit era. "We are now working on integrating all of our devices onto a microprocessor alongside transistors". "The Avalanche Photodetector achievement, which is the last in a series of prior reports from IBM Research, is the last piece of the puzzle that completes the development of the "nanophotonics toolbox" of devices necessary to build the on-chip interconnects". "With optical communications embedded into the processor chips, the prospect of building power-efficient computer systems with performance at the Exaflop level might not be a very distant future."

Nanometre

The nanometre (International spelling as used by the International Bureau of Weights and Measures; SI symbol: nm) or nanometre (American spelling) is a unit of length in the metric system, equal to one billionth of a metre. The name combines the SI prefix *nano-* (from the Ancient Greek νÜνïò, *nanos*, "dwarf") with the parent unit name *metre* (from Greek μÝôñïν, *metrïn*, "unit of measurement"). It can be written in scientific notation as 1×10 m, in engineering notation as 1 E"9 m, and is simply 1 / 1,000,000,000 m. One nanometre equals ten angstroms.

Advanced Nanotechnology Memories

For the next generation of memory to be able to keep up with the inevitable demand of new devices like the iPad, a new marriage between nanotechnology and memory devices will be vital. These super hand-held computing devices will require an inordinate amount of memory, which will need to be compact, energy efficient and fast. And so across industry and academia, the race is on to develop the next generation of device memory technology. One area that has attracted a good amount of interest is nanomemory technology. Nanomemory devices use nanoparticles and nanowires to create dense memory that is comparable or better than current memory technology in terms of functionality, while also using less energy and being faster and more flexible.

Nano-RAM

Nano-RAM is a proprietary computer memory technology from the company Nantero. It is a type of nonvolatile random access memory based on the position of carbon nanotubes deposited on a chip-like substrate. In theory, the small size of the nanotubes allows for very high density memories. Nantero also refers to it as NRAM.

Technology

The first generation Nantero NRAM technology was based on a three-terminal semiconductor device where a third terminal is used to switch the memory cell between memory states. The second generation NRAM technology is based on a two-terminal memory cell. The two-terminal cell has advantages such as a smaller cell size, better scalability to sub-20 nm nodes, and the ability to passivate the memory cell during fabrication.

In a non-woven fabric matrix of carbon nanotubes (CNTs), crossed nanotubes can either be touching or slightly separated depending on their position. When close to each other, the carbon nanotubes come under the influence of Van der Waal's forces. Each NRAM "cell" consists of an interlinked network of CNTs located between two electrodes as illustrated in Figure. The CNT fabric is located between two metal electrodes, which is defined and etched by photolithography, and forms the NRAM cell.

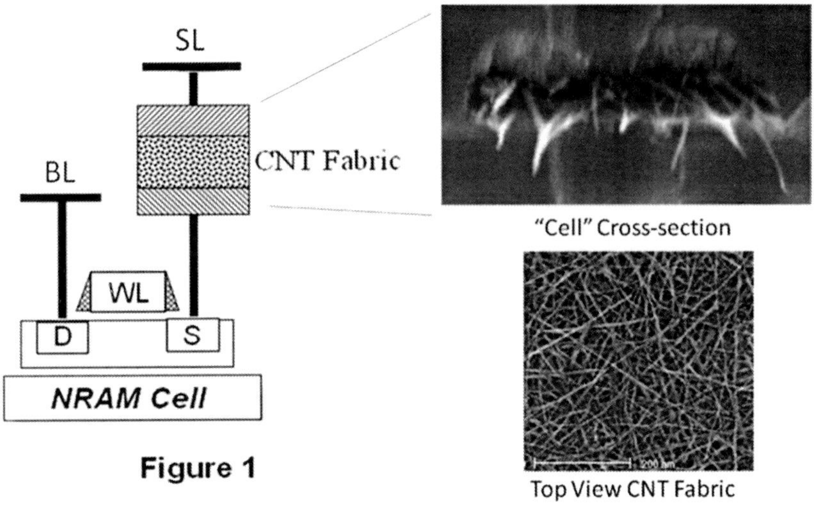

Figure: Carbon nanotube fabric

The NRAM acts as a resistive non-volatile random access memory (RAM) and can be placed in two or more resistive modes depending on the resistive state of the CNT fabric. When the CNTs are not in contact the resistance state of the fabric is high and represents an "off" or "0" state. When the CNTs are brought into contact, the resistance state of the fabric is low and represents an "on" or "1" state. NRAM acts as a memory because the two resistive states are very stable. In the 0 state, the CNTs (or a portion of them) are not in contact and remain in a separated state due to the stiffness of the CNTs resulting in a high resistance or low current measurement state between the top and bottom electrodes. In the 1 state, the CNTs (or a portion of them) are in contact and remain contacted due to Van der Waals forces between the CNTs, resulting in a low resistance or high current measurement state between the top and bottom electrodes.

To switch the NRAM between states, a small voltage greater than the read voltage is applied between top and bottom electrodes. If the NRAM is in the 0 state, the voltage applied will cause an electrostatic attraction between the CNTs close each other causing a SET operation. After the applied voltage is removed, the CNTs remain in a 1 or low resistance state due to physical adhesion (Van der Waals force) with an activation energy (E_a) of approximately 5eV. If the NRAM cell is in the 1 state, applying a voltage greater than the read voltage will generate CNT phonon excitations with sufficient energy to separate the CNT junctions. This is the phonon driven RESET operation. The CNTs remain in the OFF or high resistance state due to the high mechanical stiffness (Young's Modulus 1 TPa) with an activation energy (E_a) much greater than 5 eV. Due to the high activation energy (> 5eV) required for switching between states, the NRAM switch resists outside interference like radiation and operating temperature that can erase or flip conventional memories like DRAM.

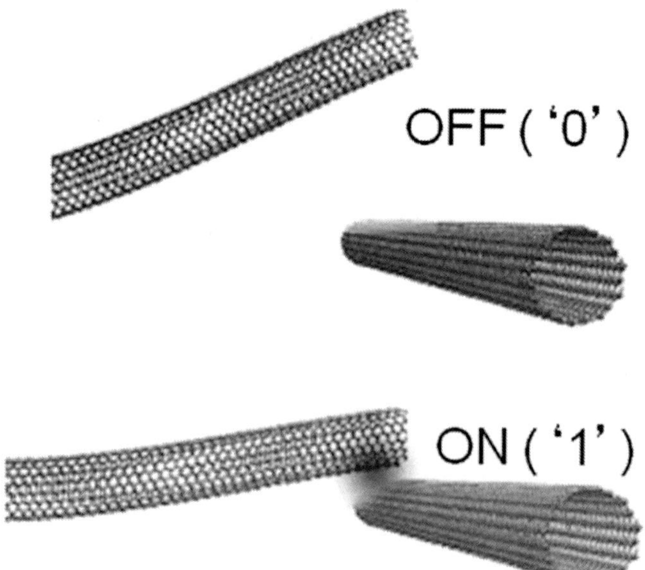

Figure: *Carbon nanotube contact points*

NRAMs are fabricated by depositing a uniform layer of CNTs onto a prefabricated array of drivers such as transistors as shown in Figure. The bottom electrode of the NRAM cell is in contact with the underlying via (electronics) connecting the cell to the driver. The bottom electrode may be fabricated as part of the underlying via or it may be fabricated simultaneously with the NRAM cell, when the cell is photolithographically defined and etched. Before the cell is photolithographically defined and etched, the top electrode is deposited as a metal film onto the CNT layer so that the top metal electrode is patterned and etched during the definition of the NRAM cell. Following the dielectric passivation and fill of the array, the top metal electrode is exposed by etching back the overlying dielectric using a smoothing process such as chemical-mechanical planarization. With the top electrode exposed, the next level of metal wiring interconnect is fabricated to complete the NRAM array. Using a cross-grid interconnect arrangement, the NRAM and driver, (the cell), forms a memory array similar to other memory arrays. A single cell can be selected by applying the proper voltages to the word line (WL), bit line (BL), and select lines (SL) without disturbing the other cells in the array.

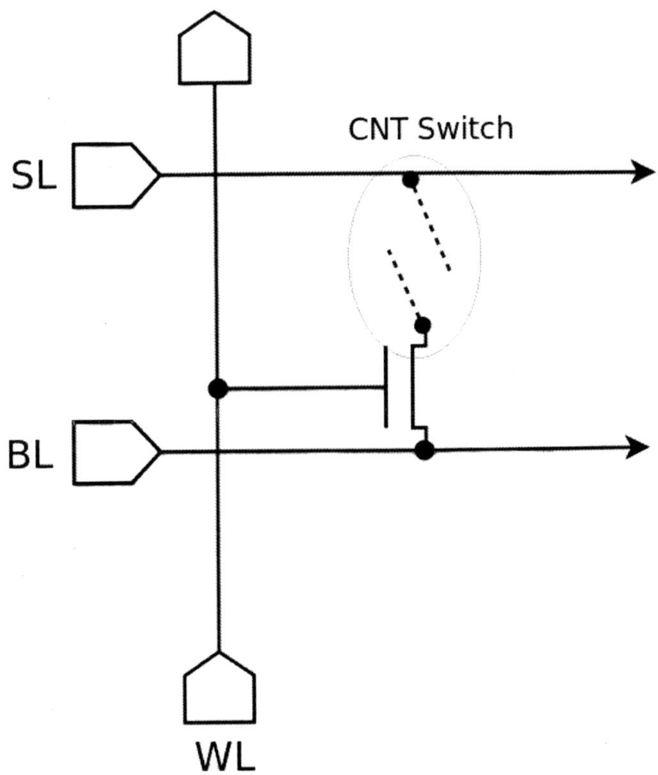

***Figure:** CNT switch*

Characteristics

NRAM has a density, at least in theory, similar to that of DRAM. DRAM includes capacitors, which are essentially two small metal plates with a thin insulator between them. NRAM has terminals and electrodes roughly the same size as the plates in a DRAM, the nanotubes between them being so much smaller they add nothing to the overall size. However it seems there is a minimum size at which a DRAM can be built, below which there is simply not enough charge being stored on the plates. NRAM appears to be limited only by lithography. This means that NRAM may be able to become much denser than DRAM, perhaps also less expensive. Unlike DRAM, NRAM does not require power to "refresh" it, and will retain its memory even after power is removed. Thus the power needed to write and retain the memory state of the device is much lower than DRAM, which has to build up charge on the cell plates. This means that NRAM might compete with DRAM in terms of cost, but also require less power, and as a result also be much faster because write performance is largely determined by the total charge needed. NRAM can theoretically reach performance similar to SRAM, which is faster than DRAM but much less dense, and thus much more expensive.

Comparison with Other Non-Volatile Memory

Compared with other non-volatile random-access memory (NVRAM) technologies, NRAM has several advantages. In flash memory, the common form of NVRAM, each cell resembles a MOSFET transistor with a control gate (CG) modulated by a floating gate (FG) interposed between the CG and the FG. The FG is surrounded by an insulating dielectric, typically an oxide. Since the FG is electrically isolated by the surrounding dielectric, any electrons placed on the FG will be trapped on the FG which screens the CG from the channel of the transistor and modifies the threshold voltage (VT) of the transistor. By writing and controlling the amount of charge placed on the FG, the FG controls the conduction state of the MOSFET flash device depending on the VT of the cell selected. The current flowing through the MOSFET channel is sensed to determine the state of the cell forming a binary code where a 1 state (current flow) when an appropriate CG voltage is applied and a 0 state (no current flow) when the CG voltage is applied.

After being written to, the insulator traps electrons on the FG, locking it into the 0 state. However, in order to change that bit, the insulator has to be "overcharged" to erase any charge already stored in it. This requires higher voltage, about 10 volts, much more than a battery can provide. Flash systems include a "charge pump" that slowly builds up power and releases it at higher voltage. This process is not only slow, but degrades the insulators. For this reason flash has a limited number of writes before the device will no longer operate effectively.

NRAM reads and writes are both "low energy" in comparison to flash (or DRAM for that matter due to "refresh"), meaning NRAM could have longer battery life. It may also be much faster to write than either, meaning it may be used to replace both. Modern phones include flash memory for storing phone numbers, DRAM for higher performance working memory because flash is too slow, and some SRAM for even higher performance. Some NRAM could be placed on the CPU to act as the CPU cache, and more in other chips replacing both the DRAM and flash.

NRAM is one of a variety of new memory systems, many of which claim to be "universal" in the same fashion as NRAM – replacing everything from flash to DRAM to SRAM.

An alternative memory ready for use is ferroelectric RAM (FRAM or FeRAM). FeRAM adds a small amount of a ferro-electric material to a DRAM cell. The state of the field in the material encodes the bit in a non-destructive format. FeRAM has advantages of NRAM, although the smallest possible cell size is much larger than for NRAM. FeRAM is used in applications where the limited number of writes of flash is an issue. FeRAM read operations are destructive, requiring a restoring write operation afterwards.

Other more speculative memory systems include magnetoresistive random-access memory (MRAM) and phase-change memory (PRAM). MRAM is based on a grid of magnetic tunnel effect junctions. MRAM's reads the memory using the tunnel magnetoresistance effect, allowing it to read the memory both non-destructively and with very little power. Early MRAM used field induced writing, reached a limit in terms of size, which kept it much larger than flash devices. However, new MRAM techniques might overcome the size limitation to make MRAM competitive even with flash memory. The techniques are Thermal Assisted Switching (TAS), developed by Crocus Technology, and Spin Torque Transfer on which Crocus, Hynix, IBM, and other companies were working in 2009.

PRAM is based on a technology similar to that in a writable CD or DVD, using a phase-change material that changes its magnetic or electrical properties instead of its optical ones. The PRAM material itself is scalable but requires a larger current source.

Carbon Nanotube

Carbon nanotubes (CNTs) are allotropes of carbon with a cylindrical nanostructure. Nanotubes have been constructed with length-to-diameter ratio of up to 132,000,000:1, significantly larger than for any other material. These cylindrical carbon molecules have unusual properties, which are valuable for nanotechnology, electronics, optics and other fields of materials science and technology. In particular, owing to their extraordinary thermal conductivity and mechanical and electrical properties, carbon nanotubes find applications as additives to various structural materials. For instance, nanotubes form a tiny portion of the material(s) in some (primarily carbon fibre) baseball bats, golf clubs, or car parts.

Nanotubes are members of the fullerene structural family. Their name is derived from their long, hollow structure with the walls formed by one-atom-thick sheets of carbon, called graphene. These sheets are rolled at specific and discrete ("chiral") angles, and the combination of the rolling angle and radius decides the nanotube properties; for example, whether the individual nanotube shell is a metal or semiconductor. Nanotubes are categorized as single-walled nanotubes (SWNTs) and multi-walled nanotubes (MWNTs). Individual nanotubes naturally align themselves into "ropes" held together by van der Waals forces, more specifically, pi-stacking.

Applied quantum chemistry, specifically, orbital hybridization best describes chemical bonding in nanotubes. The chemical bonding of nanotubes is composed entirely of sp^2 bonds, similar to those of graphite. These bonds, which are stronger than the sp^3 bonds found in alkanes and diamond, provide nanotubes with their unique strength.

Types of Carbon Nanotubes and Related Structures

Terminology

There is no consensus on some terms describing carbon nanotubes in scientific literature: both "-wall" and "-walled" are being used in combination with "single", "double", "triple" or "multi", and the letter C is often omitted in the abbreviation; for example, multi-walled carbon nanotube (MWNT).

Single-Walled

Most single-walled nanotubes (SWNT) have a diameter of close to 1 nanometre, with a tube length that can be many millions of times longer. The structure of a SWNT can be conceptualized by wrapping a one-atom-thick layer of graphite called graphene into a seamless cylinder.

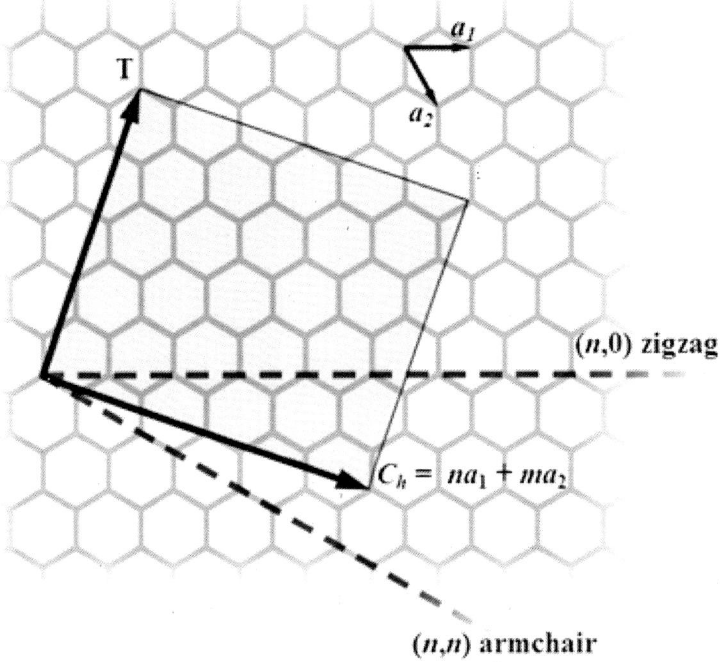

Figure: *The* (n,m) *nanotube naming scheme can be thought of as a vector* (C_h) *in an infinite graphene sheet that describes how to "roll up" the graphene sheet to make the nanotube. T denotes the tube axis, and* a_1 *and* a_2 *are the unit vectors of graphene in real space.*

The way the graphene sheet is wrapped is represented by a pair of indices (n,m). The integers n and m denote the number of unit vectors along two directions in the honeycomb crystal lattice of graphene. If $m = 0$, the nanotubes are called zigzag nanotubes, and if $n = m$, the nanotubes are called armchair nanotubes. Otherwise, they are called chiral. The diameter of an ideal nanotube can be calculated from its (n,m) indices as follows

$$d = \frac{a}{\pi}\sqrt{(n^2 + nm + m^2)} = 78.3\sqrt{((n+m)^2 - nm)}\text{pm},$$

where a = 0.246 nm.

SWNTs are an important variety of carbon nanotube because most of their properties change significantly with the (n,m) values, and this dependence is non-monotonic. In particular, their band gap can vary from zero to about 2 eV and their electrical conductivity can show metallic or semiconducting behaviour. Single-walled nanotubes are likely candidates for miniaturizing electronics. The most basic building block of these systems is the electric wire, and SWNTs with diameters of an order of a nanometre can be excellent conductors. One useful application of SWNTs is in the development of the first intermolecular field-effect transistors (FET). The first intermolecular logic gate using SWCNT FETs was made in 2001. A logic gate requires both a p-FET and an n-FET. Because SWNTs are p-FETs when exposed to oxygen and n-FETs otherwise, it is possible to protect half of an SWNT from oxygen exposure, while exposing the other half to oxygen. This results in a single SWNT that acts as a NOT logic gate with both p and n-type FETs within the same molecule.

Single-walled nanotubes are dropping precipitously in price, from around $1500 per gram as of 2000 to retail prices of around $50 per gram of as-produced 40–60% by weight SWNTs as of March 2010.

Multi-Walled

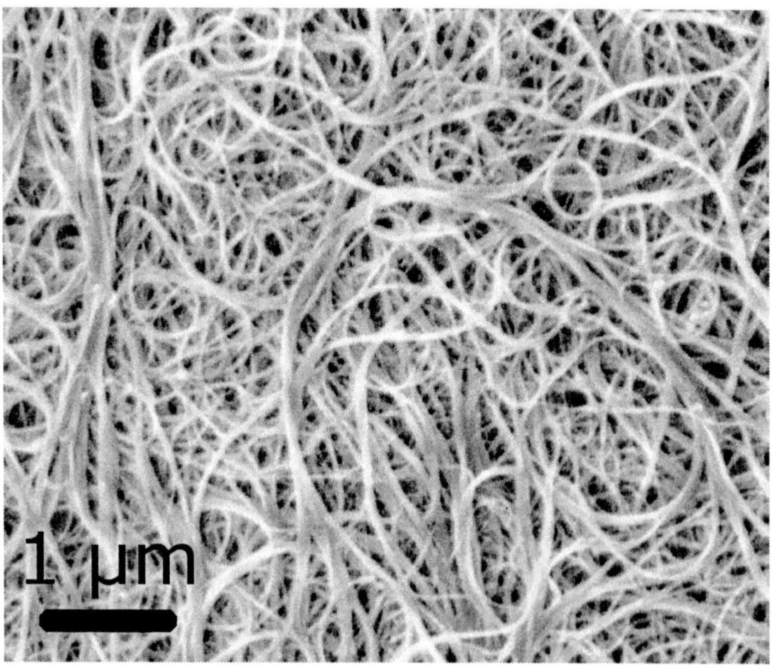

Figure: *A scanning electron microscopy image of carbon nanotubes bundles*

Multi-walled nanotubes (MWNT) consist of multiple rolled layers (concentric tubes) of graphene. There are two models that can be used to describe the structures of multi-walled nanotubes. In the *Ru Rotating model of multi-walled nanotubessian Doll* model, sheets of graphite are arranged in concentric cylinders, e.g., a (0,8) single-walled nanotube (SWNT) within a larger (0,17) single-walled nanotube. In the *Parchment* model, a single sheet of graphite is rolled in around itself, resembling a scroll of parchment or a rolled newspaper.

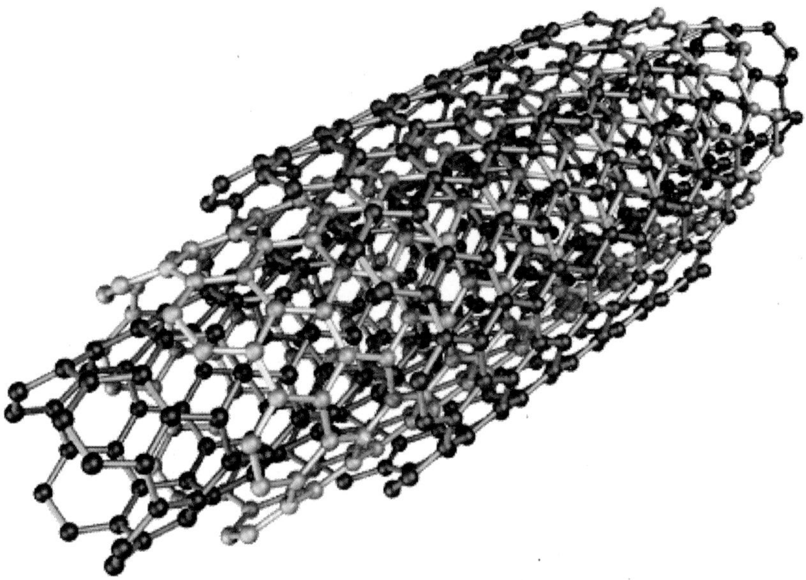

Figure: *Triple-walled armchair carbon nanotube*

The interlayer distance in multi-walled nanotubes is close to the distance between graphene layers in graphite, approximately 3.4 Å. The Russian Doll structure is observed more commonly. Its individual shells can be described as SWNTs, which can be metallic or semiconducting. Because of statistical probability and restrictions on the relative diameters of the individual tubes, one of the shells, and thus the whole MWNT, is usually a zero-gap metal.

Double-walled carbon nanotubes (DWNT) form a special class of nanotubes because their morphology and properties are similar to those of SWNT but their resistance to chemicals is significantly improved. This is especially important when functionalization is required (this means grafting of chemical functions at the surface of the nanotubes) to add new properties to the CNT. In the case of SWNT, covalent functionalization will break some C=C double bonds, leaving "holes" in the structure on the nanotube and, thus, modifying both its mechanical and electrical properties. In the case of DWNT, only the outer wall is modified. DWNT synthesis on the gram-scale was first proposed in 2003 by the CCVD technique, from the selective reduction of oxide solutions in methane and hydrogen.

The telescopic motion ability of inner shells and their unique mechanical properties will permit the use of multi-walled nanotubes as main movable arms in coming nanomechanical devices. Retraction force that occurs to telescopic motion caused by the Lennard-Jones interaction between shells and its value is about 1.5 nN.

Torus

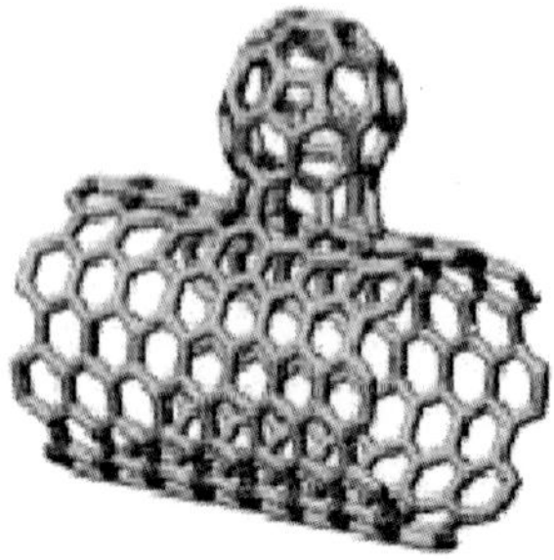

Figure: *A stable nanobud structure*

In theory, a nanotorus is a carbon nanotube bent into a torus (doughnut shape). Nanotori are predicted to have many unique properties, such as magnetic moments 1000 times larger than previously expected for certain specific radii. Properties such as magnetic moment, thermal stability, etc. vary widely depending on radius of the torus and radius of the tube.

Nanobud

Carbon nanobuds are a newly created material combining two previously discovered allotropes of carbon: carbon nanotubes and fullerenes. In this new material, fullerene-like "buds" are covalently bonded to the outer sidewalls of the underlying carbon nanotube. This hybrid material has useful properties of both fullerenes and carbon nanotubes. In particular, they have been found to be exceptionally good field emitters. In composite materials, the attached fullerene molecules may function as molecular anchors preventing slipping of the nanotubes, thus improving the composite's mechanical properties.

Graphenated Carbon Nanotubes (g-CNTs)

Graphenated CNTs are a relatively new hybrid that combines graphitic foliates grown along the sidewalls of multiwalled or bamboo style CNTs. Yu *et al.* reported on "chemically bonded graphene leaves" growing along the sidewalls of CNTs. Stoner *et al.* described these structures as "graphenated CNTs" and reported in their use for enhanced supercapacitor performance. Hsu *et al.* further reported on similar structures formed on carbon fibre paper, also for use in supercapacitor applications. The foliate density can vary as a function of deposition conditions (e.g. temperature and time) with their structure ranging from few layers of graphene (< 10) to thicker, more graphite-like.

The fundamental advantage of an integrated graphene-CNT structure is the high surface area three-dimensional framework of the

CNTs coupled with the high edge density of graphene. Graphene edges provide significantly higher charge density and reactivity than the basal plane, but they are difficult to arrange in a three-dimensional, high volume-density geometry. CNTs are readily aligned in a high density geometry (i.e., a vertically aligned forest) but lack high charge density surfaces—the sidewalls of the CNTs are similar to the basal plane of graphene and exhibit low charge density except where edge defects exist. Depositing a high density of graphene foliates along the length of aligned CNTs can significantly increase the total charge capacity per unit of nominal area as compared to other carbon nanostructures.

Nitrogen Doped Carbon Nanotubes

Nitrogen doped carbon nanotubes (N-CNT's), can be produced through 5 main methods, Chemical Vapor Deposition, high-temperature and high-pressure reactions, gas-solid reaction of amorphous carbon with NH_3 at high temperature, solid reaction, and solvothermal synthesis.

N-CNTs can also be prepared by a CVD method of pyrolysizing melamine under Ar at elevated temperatures of 800°C - 980°C. However synthesis via CVD and melamine results in the formation of bamboo structured CNTs. XPS spectra of grown N-CNT's reveals nitrogen in five main components, pyridinic nitrogen, pyrrolic nitrogen, quaternary nitrogen, and nitrogen oxides. Furthermore synthesis temperature affects the type of nitrogen configuration.

Nitrogen doping plays a pivotal role in Lithium storage. N-doping provides defects in the walls of CNT's allowing for Li ions to diffuse into interwall space. It also increases capacity by providing more favourable bind of N-doped sites. N-CNT's are also much more reactive to metal oxide nanoparticle deposition which can further enhance storage capacity, especially in anode materials for Li-ion batteries. However Boron doped nanotubes have been shown to make batteries with triple capacity.

Peapod

A Carbon peapod is a novel hybrid carbon material which traps fullerene inside a carbon nanotube. It can possess interesting magnetic properties with heating and irradiating. It can also be applied as an oscillator during theoretical investigations and predictions.

Cup-Stacked Carbon Nanotubes

Cup-stacked carbon nanotubes (CSCNTs) differ from other quasi-1D carbon structures, which normally behave as quasi-metallic conductors of electrons. CSCNTs exhibit semiconducting behaviours due to the stacking microstructure of graphene layers.

Extreme Carbon Nanotubes

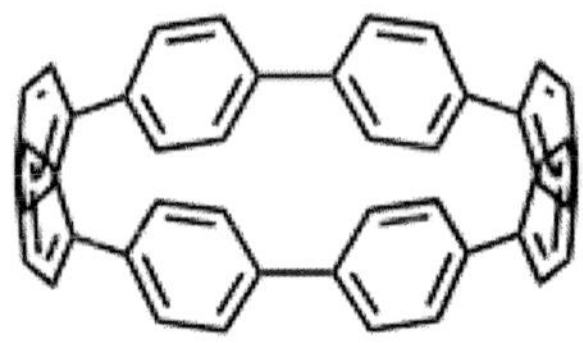

Figure: *Cycloparaphenylene*

The observation of the *longest* carbon nanotubes (18.5 cm long) was reported in 2009. These nanotubes were grown on Si substrates using an improved chemical vapor deposition (CVD) method and represent electrically uniform arrays of single-walled carbon nanotubes.

The *shortest* carbon nanotube is the organic compound cycloparaphenylene, which was synthesized in early 2009.

The *thinnest* carbon nanotube is armchair (2,2) CNT with a diameter of 3 Å. This nanotube was grown inside a multi-walled carbon nanotube. Assigning of carbon nanotube type was done by combination of high-resolution transmission electron microscopy (HRTEM), Raman spectroscopy and density functional theory (DFT) calculations.

The *thinnest freestanding* single-walled carbon nanotube is about 4.3 Å in diameter. Researchers suggested that it can be either (5,1) or (4,2) SWCNT, but exact type of carbon nanotube remains questionable. (3,3), (4,3) and (5,1) carbon nanotubes (all about 4 Å in diameter) were unambiguously identified using aberration-corrected high-resolution transmission electron microscopy inside double-walled CNTs.

The *highest density* of CNTs was achieved in 2013, grown on a conductive titanium-coated copper surface that was coated with co-catalysts cobalt and molybdenum at lower than typical temperatures of 450°C. The tubes averaged a height of 0.38 μm and a mass density of 1.6g cm–3. The material showed ohmic conductivity (lowest resistance ~22kΩ).

Properties

Strength: Carbon nanotubes are the strongest and stiffest materials yet discovered in terms of tensile strength and elastic modulus respectively. This strength results from the covalent sp^2 bonds formed between the individual carbon atoms. In 2000, a multi-walled carbon nanotube was tested to have a tensile strength of 63 gigapascals (GPa). (For illustration, this translates into the ability to endure tension of a weight equivalent to 6422 kg (14,158 lbs) on a cable with cross-section of 1 mm^2.) Further studies, such as one conducted in 2008, revealed that individual CNT shells have strengths of up to ~100 GPa, which is in agreement with quantum/atomistic models. Since carbon nanotubes have a low density for a solid of 1.3 to 1.4 g/cm^3, its specific strength of up to 48,000 kN·m·kg^{-1} is the best of known materials, compared to high-carbon steel's 154 kN·m·kg^{-1}.

Under excessive tensile strain, the tubes will undergo plastic deformation, which means the deformation is permanent. This deformation begins at strains of approximately 5% and can increase the maximum strain the tubes undergo before fracture by releasing strain energy.

Although the strength of individual CNT shells is extremely high, weak shear interactions between adjacent shells and tubes leads to significant reductions in the effective strength of multi-walled carbon nanotubes and carbon nanotube bundles down to only a few GPa's. This limitation has been recently addressed by applying high-energy electron irradiation, which crosslinks inner shells and tubes, and effectively increases the strength of these materials to ~60 GPa for multi-walled carbon nanotubes and ~17 GPa for double-walled carbon nanotube bundles.

CNTs are not nearly as strong under compression. Because of their hollow structure and high aspect ratio, they tend to undergo buckling when placed under compressive, torsional, or bending stress.

Comparison of mechanical properties			
Material	***Young's modulus (TPa)***	***Tensile strength (GPa)***	***Elongation at break (%)***
SWNT^E	~1 (from 1 to 5)	13–53	16
Armchair SWNT^T	0.94	126.2	23.1
Zigzag SWNT^T	0.94	94.5	15.6–17.5
Chiral SWNT	0.92		
MWNT^E	0.2–0.8–0.95	11–63–150	
Stainless steel^E	0.186–0.214	0.38–1.55	15–50
Kevlar–29&149^E	0.06–0.18	3.6–3.8	~2

the above discussion referred to axial properties of the nanotube, whereas simple geometrical considerations suggest that carbon nanotubes should be much softer in the radial direction than along the tube axis. Indeed, tem observation of radial elasticity suggested that even the van der waals forces can deform two adjacent nanotubes. Nanoindentation experiments, performed by several groups on multiwalled carbon nanotubes and tapping/contact mode atomic force microscope measurement performed on single-walled carbon nanotube, indicated young's modulus of the order of several gpa confirming that cnts are indeed rather soft in the radial direction.

Hardness

Standard single-walled carbon nanotubes can withstand a pressure up to 25 GPa without deformation. They then undergo a transformation to superhard phase nanotubes. Maximum pressures measured using current experimental techniques are around 55 GPa. However, these new superhard phase nanotubes collapse at an even higher, albeit unknown, pressure.

The bulk modulus of superhard phase nanotubes is 462 to 546 GPa, even higher than that of diamond (420 GPa for single diamond crystal).

Kinetic Properties

Multi-walled nanotubes are multiple concentric nanotubes precisely nested within one another. These exhibit a striking telescoping property whereby an inner nanotube core may slide, almost without friction, within its outer nanotube shell, thus creating an atomically perfect linear or rotational bearing. This is one of the first true examples of molecular nanotechnology, the precise positioning of atoms to create useful machines. Already, this property has been utilised to create the world's smallest rotational motor. Future applications such as a gigahertz mechanical oscillator are also envisioned.

Electrical Properties

because of the symmetry and unique electronic structure of graphene, the structure of a nanotube strongly affects its electrical properties. For a given (n,m) nanotube, if $n = m$, the nanotube is metallic; if $n - m$ is a multiple of 3, then the nanotube is semiconducting with a very small band gap, otherwise the nanotube is a moderate semiconductor. Thus all armchair $(n = m)$ nanotubes are metallic, and nanotubes (6,4), (9,1), etc. are semiconducting.

However, this rule has exceptions, because curvature effects in small diameter carbon nanotubes can strongly influence electrical properties. Thus, a (5,0) SWCNT that should be semiconducting in fact is metallic according to the calculations. Likewise, *vice versa*—zigzag and chiral SWCNTs with small diameters that should be metallic have finite gap (armchair nanotubes remain metallic). In theory, metallic nanotubes can carry an electric current density of 4×10^9 A/cm^2, which is more than 1,000 times greater than those of metals such as copper, where for copper interconnects current densities are limited by electromigration.

Because of their nanoscale cross-section, electrons propagate only along the tube's axis. As a result, carbon nanotubes are frequently referred to as one-dimensional conductors. The maximum electrical conductance of a single-walled carbon nanotube is $2G_0$, where $G_0 = 2e^2/h$ is the conductance of a single ballistic quantum channel. There have been reports of intrinsic superconductivity in carbon nanotubes. Many other experiments, however, found no evidence of superconductivity, and the validity of these claims of intrinsic superconductivity remains a subject of debate.

Optical Properties

Thermal Properties: All nanotubes are expected to be very good thermal conductors along the tube, exhibiting a property known as "ballistic conduction", but good insulators laterally to the tube axis. Measurements show that a SWNT has a room-temperature thermal conductivity along its axis of about 3500 $W \cdot m^{-1} \cdot K^{-1}$; compare this to copper, a metal well known for its good thermal conductivity, which transmits 385 $W \cdot m^{-1} \cdot K^{-1}$.

A SWNT has a room-temperature thermal conductivity across its axis (in the radial direction) of about 1.52 $W \cdot m^{-1} \cdot K^{-1}$, which is about as thermally conductive as soil. The temperature stability of carbon nanotubes is estimated to be up to 2800 °C in vacuum and about 750 °C in air.

Defects

As with any material, the existence of a crystallographic defect affects the material properties. Defects can occur in the form of atomic vacancies. High levels of such defects can lower the tensile strength by up to 85%. An important example is the Stone Wales defect, which creates a pentagon and heptagon pair by rearrangement of the bonds. Because of the very small structure of CNTs, the tensile strength of the tube is dependent on its weakest segment in a similar manner to a chain, where the strength of the weakest link becomes the maximum strength of the chain.

Crystallographic defects also affect the tube's electrical properties. A common result is lowered conductivity through the defective region of the tube. A defect in armchair-type tubes (which can conduct electricity) can cause the surrounding region to become semiconducting, and single monatomic vacancies induce magnetic properties.

Crystallographic defects strongly affect the tube's thermal properties. Such defects lead to phonon scattering, which in turn increases the relaxation rate of the phonons. This reduces the mean free path and reduces the thermal conductivity of nanotube structures. Phonon transport simulations indicate that substitutional defects such as nitrogen or boron will primarily lead to scattering of high-frequency optical phonons. However, larger-scale defects such as Stone Wales defects cause phonon scattering over a wide range of frequencies, leading to a greater reduction in thermal conductivity.

Toxicity

The toxicity of carbon nanotubes has been an important question in nanotechnology. As of 2007, such research has just begun. The data are still fragmentary and subject to criticism. Preliminary results highlight the difficulties in evaluating the toxicity of this heterogeneous material. Parameters such as structure, size distribution, surface area, surface chemistry, surface charge, and agglomeration state as well as purity of the samples, have considerable impact on the reactivity of carbon nanotubes. However, available data clearly show that, under some conditions, nanotubes can cross membrane barriers, which suggests that, if raw materials reach the organs, they can induce harmful effects such as inflammatory and fibrotic reactions.

Under certain conditions CNTs can enter human cells and accumulate in the cytoplasm, causing cell death.

Results of rodent studies collectively show that regardless of the process by which CNTs were synthesized and the types and amounts of metals they contained, CNTs were capable of producing inflammation, epithelioid granulomas (microscopic nodules), fibrosis, and biochemical/ toxicological changes in the lungs. Comparative toxicity studies in which mice were given equal weights of test materials showed that SWCNTs were more toxic than quartz, which is considered a serious occupational health hazard when chronically inhaled. As a control, ultrafine carbon black was shown to produce minimal lung responses.

Carbon nano tubes deposit in the alveolar ducts by aligning length wise with the airways; the nano tubes will often combine with metals. The needle-like fibre shape of CNTs is similar to asbestos fibres. This raises the idea that widespread use of carbon nanotubes may lead to pleural mesothelioma, a cancer of the lining of the lungs or peritoneal mesothelioma, a cancer of the lining of the abdomen (both caused by exposure to asbestos). A recently published pilot study supports this prediction. Scientists exposed the mesothelial lining of the body cavity of mice to long multiwalled carbon nanotubes and observed asbestos-like, length-dependent, pathogenic behaviour that included inflammation and formation of lesions known as granulomas. Authors of the study conclude:

This is of considerable importance, because research and business communities continue to invest heavily in carbon nanotubes for a wide range of products under the assumption that they are no more hazardous than graphite. Our results suggest the need for further research and great caution before introducing such products into the market if long-term harm is to be avoided.

Although further research is required, the available data suggests that under certain conditions, especially those involving chronic exposure, carbon nanotubes can pose a serious risk to human health.

Synthesis

Figure: *powder of carbon nanotubes*

techniques have been developed to produce nanotubes in sizable quantities, including arc discharge, laser ablation, high-pressure carbon monoxide disproportionation, and chemical vapor deposition (cvd). Most of these processes take place in vacuum or with process gases. cvd growth of cnts can occur in vacuum or at atmospheric pressure. Large quantities of nanotubes can be synthesized by these methods; advances in catalysis and continuous growth are making cnts more commercially viable.

Arc Discharge

Nanotubes were observed in 1991 in the carbon soot of graphite electrodes during an arc discharge, by using a current of 100 amps, that was intended to produce fullerenes. However the first macroscopic production of carbon nanotubes was made in 1992 by two researchers at NEC's Fundamental Research Laboratory. The method used was the same as in 1991. During this process, the carbon contained in the negative electrode sublimates because of the high-discharge temperatures. Because nanotubes were initially discovered using this technique, it has been the most widely used method of nanotube synthesis.

The yield for this method is up to 30% by weight and it produces both single- and multi-walled nanotubes with lengths of up to 50 micrometres with few structural defects.

Laser Ablation

In laser ablation, a pulsed laser vaporizes a graphite target in a high-temperature reactor while an inert gas is bled into the chamber. Nanotubes develop on the cooler surfaces of the reactor as the vaporized carbon condenses. A water-cooled surface may be included in the system to collect the nanotubes.

This process was developed by Dr. Richard Smalley and co-workers at Rice University, who at the time of the discovery of carbon nanotubes, were blasting metals with a laser to produce various metal molecules. When they heard of the existence of nanotubes they replaced the metals with graphite to create multi-walled carbon nanotubes. Later that year the team used a composite of graphite and metal catalyst particles (the best yield was from a cobalt and nickel mixture) to synthesize single-walled carbon nanotubes.

The laser ablation method yields around 70% and produces primarily single-walled carbon nanotubes with a controllable diameter determined by the reaction temperature. However, it is more expensive than either arc discharge or chemical vapor deposition.

Plasma Torch

Single-walled carbon nanotubes can also be synthesized by a thermal plasma method. It was first invented in 2000 at INRS (Institut National de la Recherche Scientifique in Varennes, Canada), by Olivier Smiljanic. In this method, the aim is to reproduce the conditions prevailing in the arc discharge and laser ablation approaches, but a carbon-containing gas is used instead of graphite vapors to supply the carbon necessary for the production of SWNT. Doing so, the growth of SWNT is more efficient (decomposing a carbon containing gas can be 10 times less energy consuming than graphite vaporization). It is also continuous and occurs at low cost. To produce a continuous process, a gas mixture composed of argon, ethylene and ferrocene is introduced into a microwave plasma torch, where it is atomized by the atmospheric pressure plasma, which has the form of an intense 'flame'. The fumes created by the flame are found to contain SWNT, metallic and carbon nanoparticles and amorphous carbon.

Another way to produce single-walled carbon nanotubes with a plasma torch, is to use the induction thermal plasma method, implemented in 2005 by groups from the University of Sherbrooke and the National Research Council of Canada. The method is similar to arc-discharge in that both use ionized gas to reach the high temperature necessary to vaporize carbon-containing substances and the metal catalysts necessary for the ensuing nanotube growth. The thermal plasma is induced by high frequency oscillating currents in a coil, and is maintained in flowing inert gas. Typically, a feedstock of carbon black and metal catalyst particles is fed into the plasma, and then cooled down to form single-walled carbon nanotubes. Different single-wall carbon nanotube diameter distributions can be synthesized.

The induction thermal plasma method can produce up to 2 grams of nanotube material per minute, which is higher than the arc-discharge or the laser ablation methods.

Chemical Vapor Deposition (CVD)

The catalytic vapor phase deposition of carbon was reported in 1952 and 1959, but it was not until 1993 that carbon nanotubes were formed by this process. In 2007, researchers at the University of Cincinnati (UC) developed a process to grow aligned carbon nanotube arrays of 18 mm length on a FirstNano ET3000 carbon nanotube growth system.

During CVD, a substrate is prepared with a layer of metal catalyst particles, most commonly nickel, cobalt, iron, or a combination. The metal nanoparticles can also be produced by other ways, including reduction of oxides or oxides solid solutions. The diameters of the nanotubes that are to be grown are related to the size of the metal particles. This can be controlled by patterned (or masked) deposition of the metal, annealing, or by plasma etching of a metal layer. The substrate is heated to approximately 700°C.

To initiate the growth of nanotubes, two gases are bled into the reactor: a process gas (such as ammonia, nitrogen or hydrogen) and a carbon-containing gas (such as acetylene, ethylene, ethanol or methane). Nanotubes grow at the sites of the metal catalyst; the carbon-containing gas is broken apart at the surface of the catalyst particle, and the carbon is transported to the edges of the particle, where it forms the nanotubes. This mechanism is still being studied. The catalyst particles can stay at the tips of the growing nanotube during growth, or remain at the nanotube base, depending on the adhesion between the catalyst particle and the substrate. Thermal catalytic decomposition of hydrocarbon has become an active area of research and can be a promising route for the bulk production of CNTs. Fluidised bed reactor is the most widely used reactor for CNT preparation. Scale-up of the reactor is the major challenge.

CVD is a common method for the commercial production of carbon nanotubes. For this purpose, the metal nanoparticles are mixed with a catalyst support such as MgO or Al_2O_3 to increase the surface area for higher yield of the catalytic reaction of the carbon feedstock with the metal particles. One issue in this synthesis route is the removal of the catalyst support via an acid treatment, which sometimes could destroy the original structure of the carbon nanotubes. However, alternative catalyst supports that are soluble in water have proven effective for nanotube growth.

If a plasma is generated by the application of a strong electric field during growth (plasma-enhanced chemical vapor deposition), then the nanotube growth will follow the direction of the electric field. By adjusting the geometry of the reactor it is possible to synthesize vertically aligned carbon nanotubes (i.e., perpendicular to the substrate), a morphology that has been of interest to researchers interested in the electron emission from nanotubes. Without the plasma, the resulting nanotubes are often randomly oriented. Under certain reaction conditions, even in the absence of a plasma, closely spaced nanotubes will maintain a vertical growth direction resulting in a dense array of tubes resembling a carpet or forest.

Of the various means for nanotube synthesis, CVD shows the most promise for industrial-scale deposition, because of its price/unit ratio, and because CVD is capable of growing nanotubes directly on a desired substrate, whereas the nanotubes must be collected in the other growth techniques. The growth sites are controllable by careful deposition of the catalyst. In 2007, a team from Meijo University demonstrated a high-efficiency CVD technique for growing carbon nanotubes from camphor. Researchers at Rice University, until recently led by the late Richard Smalley, have concentrated upon finding methods to produce large, pure amounts of particular types of nanotubes. Their approach grows long fibres from many small seeds cut from a single nanotube; all of the resulting fibres were found to be of the same diameter as the original nanotube and are expected to be of the same type as the original nanotube.

Super-Growth CVD

Super-growth CVD (water-assisted chemical vapor deposition) was developed by Kenji Hata, Sumio Iijima and co-workers at AIST, Japan.

In this process, the activity and lifetime of the catalyst are enhanced by addition of water into the CVD reactor. Dense millimetre-tall nanotube "forests", aligned normal to the substrate, were produced. The forests growth rate could be expressed, as

$$H(t) = \beta\tau_o(1 - e^{-t/\tau_o|last1=Smith|first1=Brian W.|journal=Nature|volume=396|issue=6709|pages=323|last2=Monthioux|first2=Marc|last3=Luzzi|first3=David E.}).$$

In this equation, β is the initial growth rate and τ_o is the characteristic catalyst lifetime.

Their specific surface exceeds 1,000 m^2/g (capped) or 2,200 m^2/g (uncapped), surpassing the value of 400–1,000 m^2/g for HiPco samples. The synthesis efficiency is about 100 times higher than for the laser ablation method. The time required to make SWNT forests of the height of 2.5 mm by this method was 10 minutes in 2004. Those SWNT forests can be easily separated from the catalyst, yielding clean SWNT material (purity >99.98%) without further purification. For comparison, the as-grown HiPco CNTs contain about 5–35% of metal impurities; it is therefore purified through dispersion and centrifugation that damages the nanotubes. Super-growth avoids this problem. Patterned highly organised single-walled nanotube structures were successfully fabricated using the super-growth technique.

The mass density of super-growth CNTs is about 0.037 g/cm^3. It is much lower than that of conventional CNT powders (~1.34 g/cm^3), probably because the latter contain metals and amorphous carbon.

The super-growth method is basically a variation of CVD. Therefore, it is possible to grow material containing SWNT, DWNTs and MWNTs, and to alter their ratios by tuning the growth conditions. Their ratios change by the thinness of the catalyst. Many MWNTs are included so that the diameter of the tube is wide.

The vertically aligned nanotube forests originate from a "zipping effect" when they are immersed in a solvent and dried. The zipping effect is caused by the surface tension of the solvent and the van der Waals forces between the carbon nanotubes. It aligns the nanotubes into a dense material, which can be formed in various shapes, such as sheets and bars, by applying weak compression during the process. Densification increases the Vickers hardness by about 70 times and density is 0.55 g/cm^3. The packed carbon nanotubes are more than 1 mm long and have a carbon purity of 99.9% or higher; they also retain the desirable alignment properties of the nanotubes forest.

Natural, Incidental, and Controlled Flame Environments

Fullerenes and carbon nanotubes are not necessarily products of high-tech laboratories; they are commonly formed in such mundane places as ordinary flames, produced by burning methane, ethylene, and benzene, and they have been found in soot from both indoor and outdoor air. However, these naturally occurring varieties can be highly irregular in size and quality because the environment in which they are produced is often highly uncontrolled. Thus, although they can be used in some applications, they can lack in the high degree of uniformity necessary to satisfy the many needs of both research and industry. Recent efforts have focused on producing more uniform carbon nanotubes in controlled flame environments. Such methods have promise for large-scale, low-cost nanotube synthesis based on theoretical models, though they must compete with rapidly developing large scale CVD production.

Removal of Catalysts

Nanoscale metal catalysts are important ingredients for fixed- and fluidized-bed CVD synthesis of CNTs. They allow increasing the growth efficiency of CNTs and may give control over their structure and chirality. During synthesis, catalysts can convert carbon precursors into tubular carbon structures but can also form encapsulating carbon overcoats. Together with metal oxide supports they may therefore attach to or become incorporated into the CNT product. The presence of metal impurities can be problematic for many applications. Especially catalyst metals like nickel, cobalt or yttrium may be of toxicological concern. While un-encapsulated catalyst metals may be readily removable by acid washing, encapsulated ones require oxidative treatment for opening their carbon shell. The effective removal of catalysts, especially of encapsulated ones, while preserving the CNT structure is a challenge and has been addressed in many studies. A new approach to break carbonaceaous catalyst encapsulations is based on rapid thermal annealing.

Application-Related Issues

Many electronic applications of carbon nanotubes crucially rely on techniques of selectively producing either semiconducting or metallic CNTs, preferably of a certain chirality. Several methods of separating semiconducting and metallic CNTs are known, but most of them are not yet suitable for large-scale technological processes. The most efficient method relies on density-gradient ultracentrifugation, which separates surfactant-wrapped nanotubes by the minute difference in their density. This density difference often translates into difference in the nanotube diameter and (semi)conducting properties. Another method of separation uses a sequence of freezing, thawing, and compression of SWNTs embedded in agarose gel. This process results in a solution containing 70% metallic SWNTs and leaves a gel containing 95% semiconducting

SWNTs. The diluted solutions separated by this method show various colours. The separated carbon nanotubes using this method have been applied to electrodes, e.g. electric double-layer capacitor. Moreover, SWNTs can be separated by the column chromatography method. Yield is 95% in semiconductor type SWNT and 90% in metallic type SWNT.

In addition to separation of semiconducting and metallic SWNTs, it is possible to sort SWNTs by length, diameter, and chirality. The highest resolution length sorting, with length variation of <10%, has thus far been achieved by size exclusion chromatography (SEC) of DNA-dispersed carbon nanotubes (DNA-SWNT). SWNT diameter separation has been achieved by density-gradient ultracentrifugation (DGU) using surfactant-dispersed SWNTs and by ion-exchange chromatography (IEC) for DNA-SWNT. Purification of individual chiralities has also been demonstrated with IEC of DNA-SWNT: specific short DNA oligomers can be used to isolate individual SWNT chiralities. Thus far, 12 chiralities have been isolated at purities ranging from 70% for (8,3) and (9,5) SWNTs to 90% for (6,5), (7,5) and (10,5)SWNTs. There have been successful efforts to integrate these purified nanotubes into devices, e. g. FETs.

An alternative to separation is development of a selective growth of semiconducting or metallic CNTs. Recently, a new CVD recipe that involves a combination of ethanol and methanol gases and quartz substrates resulting in horizontally aligned arrays of 95–98% semiconducting nanotubes was announced.

Nanotubes are usually grown on nanoparticles of magnetic metal (Fe, Co), which facilitates production of electronic (spintronic) devices. In particular, control of current through a field-effect transistor by magnetic field has been demonstrated in such a single-tube nanostructure.

Current Applications

Current use and application of nanotubes has mostly been limited to the use of bulk nanotubes, which is a mass of rather unorganised fragments of nanotubes. Bulk nanotube materials may never achieve a tensile strength similar to that of individual tubes, but such composites may, nevertheless, yield strengths sufficient for many applications. Bulk carbon nanotubes have already been used as composite fibres in polymers to improve the mechanical, thermal and electrical properties of the bulk product.

- Easton-Bell Sports, Inc. have been in partnership with Zyvex Performance Materials, using CNT technology in a number of their bicycle components—including flat and riser handlebars, cranks, forks, seatposts, stems and aero bars.

- Zyvex Technologies has also built a 54' maritime vessel, the Piranha Unmanned Surface Vessel, as a technology demonstrator for what is possible using CNT technology. CNTs help improve the structural performance of the vessel, resulting in a lightweight 8,000 lb boat that can carry a payload of 15,000 lb over a range of 2,500 miles.
- Amroy Europe Oy manufactures Hybtonite carbon nanoepoxy resins where carbon nanotubes have been chemically activated to bond to epoxy, resulting in a composite material that is 20% to 30% stronger than other composite materials. It has been used for wind turbines, marine paints and variety of sports gear such as skis, ice hockey sticks, baseball bats, hunting arrows, and surfboards.

Other current applications include:

- tips for atomic force microscope probes
- in tissue engineering, carbon nanotubes can act as scaffolding for bone growth

Potential Applications

The strength and flexibility of carbon nanotubes makes them of potential use in controlling other nanoscale structures, which suggests they will have an important role in nanotechnology engineering. The highest tensile strength of an individual multi-walled carbon nanotube has been tested to be is 63 GPa. Carbon nanotubes were found in Damascus steel from the 17th century, possibly helping to account for the legendary strength of the swords made of it.

Structural

Because of the carbon nanotube's superior mechanical properties, many structures have been proposed ranging from everyday items like clothes and sports gear to combat jackets and space elevators. However, the space elevator will require further efforts in refining carbon nanotube technology, as the practical tensile strength of carbon nanotubes can still be greatly improved.

For perspective, outstanding breakthroughs have already been made. Pioneering work led by Ray H. Baughman at the NanoTech Institute has shown that single and multi-walled nanotubes can produce materials with toughness unmatched in the man-made and natural worlds.

- Carbon nanotubes are also a promising material as building blocks in bio-mimetic hierarchical composite materials given their exceptional mechanical properties (~1 TPa in modulus, and ~100 GPa in strength). Initial attempts to incorporate CNTs into hierarchical structures led to mechanical properties that were significantly lower than these achievable limits. Windle *et al.* have used an *in situ* chemical vapor deposition (CVD) spinning method to produce continuous CNT yarns from CVD grown CNT aerogels. With this technology, they fabricated CNT yarns with strengths as high as ~9 GPa at small gage lengths of ~1 mm, however, defects resulted in a reduction of specific strength to ~1 GPa at 20 mm gage length. Espinosa *et al.* developed high performance DWNT-polymer composite yarns by twisting and stretching ribbons of randomly oriented bundles of DWNTs thinly coated with polymeric organic compounds. These DWNT-polymer yarns exhibited unusually high energy to failure of ~100 $J \cdot g^1$ (comparable to one of the toughest natural materials – spider silk), and strength as high as ~1.4 GPa. Effort is ongoing to produce CNT composites that incorporate tougher matrix materials, such as Kevlar, to further improve on the mechanical properties toward those of individual CNTs.

Because of the high mechanical strength of carbon nanotubes, research is being made into weaving them into clothes to create stab-proof and bulletproof clothing. The nanotubes would effectively stop the bullet from penetrating the body, although the bullet's kinetic energy would likely cause broken bones and internal bleeding.

Electrical Circuits

Nanotube-based transistors, also known as carbon nanotube field-effect transistors (CNTFETs), have been made that operate at room temperature and that are capable of digital switching using a single electron. However, one major obstacle to realisation of nanotubes has been the lack of technology for mass production. In 2001 IBM researchers demonstrated how metallic nanotubes can be destroyed, leaving semiconducting ones behind for use as transistors. Their process is called "constructive destruction," which includes the automatic destruction of defective nanotubes on the wafer. This process, however, only gives control over the electrical properties on a statistical scale.

The potential of carbon nanotubes was demonstrated in 2003 when room-temperature ballistic transistors with ohmic metal contacts and high-k gate dielectric were reported, showing 20–30x higher ON current than state-of-the-art Si MOSFETs. This presented an important advance in the field as CNT was shown to potentially outperform Si. At the time, a major challenge was ohmic metal contact formation. In this regard, palladium, which is a high-work function metal was shown to exhibit Schottky barrier-free contacts to semiconducting nanotubes with diameters >1.7 nm.

The first nanotube integrated memory circuit was made in 2004. One of the main challenges has been regulating the conductivity of nanotubes. Depending on subtle surface features a nanotube may act as a plain conductor or as a semiconductor. A fully automated method has however been developed to remove non-semiconductor tubes.

Another way to make carbon nanotube transistors has been to use random networks of them. By doing so one averages all of their electrical differences and one can produce devices in large scale at the wafer level. This approach was first patented by Nanomix Inc. (date of original application June 2002). It was first published in the academic literature by the United States Naval Research Laboratory in 2003 through independent research work. This approach also enabled Nanomix to make the first transistor on a flexible and transparent substrate.

Large structures of carbon nanotubes can be used for thermal management of electronic circuits. An approximately 1 mm–thick carbon nanotube layer was used as a special material to fabricate coolers, this material has very low density, ~20 times lower weight than a similar copper structure, while the cooling properties are similar for the two materials. In 2013, researchers demonstrated a Turing-complete prototype micrometre-scale computer. Carbon nanotube transistors as logic-gate circuits with densities comparable to modern CMOS technology has not yet been demonstrated.

Electrical Cables and Wires

Wires for carrying electrical current may be fabricated from pure nanotubes and nanotube-polymer composites. It has already been demonstrated that carbon nanotube wires can successfully be used for power or data transmission. Recently small wires have been fabricated with specific conductivity exceeding copper and aluminum; these cables are the highest conductivity carbon nanotube and also highest conductivity non-metal cables. Recently, composite of carbon nanotube and copper have been shown to exhibit nearly one hundred times higher current-carrying-capacity than pure copper or gold. Significantly, the electrical conductivity of such a composite is similar to pure Cu. Thus, this Carbon nanotube-copper (CNT-Cu) composite possesses the highest observed current-carrying capacity among electrical conductors. Thus for a given cross-section of electrical conductor, the CNT-Cu composite can withstand and transport one hundred times higher current compared to metals such as copper and gold.

Actuators

The exceptional electrical and mechanical properties of carbon nanotubes have made them alternatives to the traditional electrical actuators for both microscopic and macroscopic applications. Carbon nanotubes are very good conductors of both electricity and heat, and they are also very strong and elastic molecules in certain directions.

Paper Batteries

A paper battery is a battery engineered to use a paper-thin sheet of cellulose (which is the major constituent of regular paper, among other things) infused with aligned carbon nanotubes. The nanotubes act as electrodes; allowing the storage devices to conduct electricity. The battery, which functions as both a lithium-ion battery and a supercapacitor, can provide a long, steady power output comparable to a conventional battery, as well as a supercapacitor's quick burst of high power—and while a conventional battery contains a number of separate components, the paper battery integrates all of the battery components in a single structure, making it more energy efficient.

Solar Cells

One of the promising applications of single-walled carbon nanotubes (SWNTs) is their use in solar panels, due to their strong UV/Vis-NIR absorption characteristics. Research has shown that they can provide a sizeable increase in efficiency, even at their current unoptimized state. Solar cells developed at the New Jersey Institute of Technology use a carbon nanotube complex, formed by a mixture of carbon nanotubes and carbon buckyballs (known as fullerenes) to form snake-like structures. Buckyballs trap electrons, but they can't make electrons flow. Add sunlight to excite the polymers, and the buckyballs will grab the electrons. Nanotubes, behaving like copper wires, will then be able to make the electrons or current flow.

Additional research has been conducted on creating SWNT hybrid solar panels to increase the efficiency further. These hybrids are created by combining SWNT's with photoexcitable electron donors to increase the number of electrons generated. It has been found that the interaction between the photoexcited porphyrin and SWNT generates electro-hole pairs at the SWNT surfaces. This phenomenon has been observed experimentally, and contributes practically to an increase in efficiency up to 8.5%.

Hydrogen Storage

In addition to being able to store electrical energy, there has been some research in using carbon nanotubes to store hydrogen to be used as a fuel source. By taking advantage of the capillary effects of the small carbon nanotubes, it is possible to condense gases in high density inside single-walled nanotubes. This allows for gases, most notably hydrogen (H_2), to be stored at high densities without being condensed into a liquid. Potentially, this storage method could be used on vehicles in place of gas fuel tanks for a hydrogen-powered car. A current issue regarding hydrogen-powered vehicles is the onboard storage of the fuel. Current storage methods involve cooling and condensing the H_2 gas to a liquid state for storage which causes a loss of potential energy (25–45%) when compared to the energy associated with the gaseous state. Storage using SWNTs would allow one to keep the H2 in its gaseous state, thereby increasing the storage efficiency. This method allows for a volume to energy ratio slightly smaller to that of current gas powered vehicles, allowing for a slightly lower but comparable range.

An area of controversy and frequent experimentation regarding the storage of hydrogen by adsorption in carbon nanotubes is the efficiency by which this process occurs. The effectiveness of hydrogen storage is integral to its use as a primary fuel source since hydrogen only contains about one fourth the energy per unit volume as gasoline.

Experimental Capacity

One experiment sought to determine the amount of hydrogen stored in CNTs by utilising elastic recoil detection analysis (ERDA). CNTs (primarily SWNTs) were synthesized via chemical vapor disposition (CVD) and subjected to a two-stage purification process including air oxidation and acid treatment, then formed into flat, uniform discs and exposed to pure, pressurized hydrogen at various temperatures. When the data was analyzed, it was found that the ability of CNTs to store hydrogen decreased as temperature increased. Moreover, the highest hydrogen concentration measured was ~0.18%; significantly lower than commercially viable hydrogen storage needs to be. A separate experimental work performed by using a gravimetric method also revealed the maximum hydrogen uptake capacity of CNTs to be as low as 0.2%.

In another experiment, CNTs were synthesized via CVD and their structure was characterized using Raman spectroscopy. Utilising microwave digestion, the samples were exposed to different acid concentrations and different temperatures for various amounts of time in an attempt to find the optimum purification method for SWNTs of the diameter determined earlier. The purified samples were then exposed to hydrogen gas at various high pressures, and their adsorption by weight percent was plotted. The data showed that hydrogen adsorption levels of up to 3.7% are possible with a very pure sample and under the proper conditions. It is thought that microwave digestion helps improve the hydrogen adsorption capacity of the CNTs by opening up the ends, allowing access to the inner cavities of the nanotubes.

Limitations on Efficient Hydrogen Adsorption

The biggest obstacle to efficient hydrogen storage using CNTs is the purity of the nanotubes. To achieve maximum hydrogen adsorption, there must be minimum graphene, amorphous carbon, and metallic deposits in the nanotube sample. Current methods of CNT synthesis require a purification step. However, even with pure nanotubes, the adsorption capacity is only maximized under high pressures, which are undesirable in commercial fuel tanks.

Supercapacitor

MIT Research Laboratory of Electronics uses nanotubes to improve supercapacitors. The activated charcoal used in conventional ultracapacitors has many small hollow spaces of various size, which create together a large surface to store electric charge. But as charge is quantized into elementary charges, i.e. electrons, and each such elementary charge needs a minimum space, a significant fraction of the electrode surface is not available for storage because the hollow spaces are not compatible with the charge's requirements. With a nanotube electrode the spaces may be tailored to size—few too large or too small—and consequently the capacity should be increased considerably.

Radar Absorption

Radars work in the microwave frequency range, which can be absorbed by MWNTs. Applying the MWNTs to the aircraft would cause the radar to be absorbed and therefore seem to have a smaller signature. One such application could be to paint the nanotubes onto the plane. Recently there has been some work done at the University of Michigan regarding carbon nanotubes usefulness as stealth technology on aircraft. It has been found that in addition to the radar absorbing properties, the nanotubes neither reflect nor scatter visible light, making it essentially invisible at night, much like painting current stealth aircraft black except much more effective. Current limitations in manufacturing, however, mean that current production of nanotube-coated aircraft is not possible. One theory to overcome these current limitations is to cover small particles with the nanotubes and suspend the nanotube-covered particles in a medium such as paint, which can then be applied to a surface, like a stealth aircraft.

Medical

In the Kanzius cancer therapy, single-walled carbon nanotubes are inserted around cancerous cells, then excited with radio waves, which causes them to heat up and kill the surrounding cells. Researchers at Rice University, Radboud University Nijmegen Medical Centre and University of California, Riverside have shown that carbon nanotubes and their polymer nanocomposites are suitable scaffold materials for bone cell proliferation and bone formation.

In November 2012 researchers at the American National Institute of Standards and Technology (NIST) proved that single-wall carbon nanotubes may help protect DNA molecules from damage by oxidation. (PhysOrg) (*Small*)

Textile

The previous studies on the use of CNTs for textile functionalization were focused on fibre spinning for improving physical and mechanical properties. Recently a great deal of attention has been focused on coating CNTs on textile fabrics. Various methods have been employed for modifying fabrics using CNTs. Shim et al. produced intelligent e-textiles for Human Biomonitoring using a polyelectrolyte-based coating with CNTs. Additionally, Panhuis et al. dyed textile material by immersion in either a poly (2-methoxy aniline-5-sulfonic acid) PMAS polymer solution or PMAS-SWNT dispersion with enhanced conductivity and capacitance with a durable behaviour. In another study, Hu and coworkers coated single-walled carbon nanotubes with a simple “dipping and drying” process for wearable electronics and energy storage applications. CNTs have an aligned nanotube structure and a negative surface charge. Therefore, they have similar structures to direct dyes, so the exhaustion method is applied for coating and absorbing CNTs on the fibre surface for preparing multifunctional fabric including antibacterial, electric conductive, flame retardant and electromagnetic absorbance properties.

Optical Power Detectors

A spray-on mixture of carbon nanotubes and ceramic demonstrates unprecedented ability to resist damage while absorbing laser light. Such coatings that absorb as the energy of high-powered lasers without breaking down are essential for optical power detectors that measure the output of such lasers. These are used, for example, in military equipment for defusing unexploded mines. The composite consists of multiwall carbon nanotubes and a ceramic made of silicon, carbon and nitrogen. Including boron boosts the breakdown temperature. The nanotubes and graphene-like carbon transmit heat well, while the oxidation-resistant ceramic boosts damage resistance. Creating the coating involves dispersing he nanotubes in toluene, to which a clear liquid polymer containing boron was added. The mixture was heated to 1,100 °C (2,010 °F). The result is crushed into a fine powder, dispersed again in toluene and sprayed in a thin coat on a copper surface. The coating absorbed 97.5 percent of the light from a far-infrared laser and tolerated 15 kilowatts per square centimetre for 10 seconds. Damage tolerance is about 50 percent higher than for similar coatings, e.g., nanotubes alone and carbon paint.

Loud Speaker and Earphone

Carbon nanotubes have also been applied in the acoustics(such as loudspeaker and earphone). In 2008 it was shown that a sheet of nanotubes can operate as a loudspeaker if an alternating current is applied. The sound is not produced through vibration but thermoacoustically. In 2013, a carbon nanotube (CNT) thin yarn thermoacoustic earphone together with CNT thin yarn thermoacoustic chip was demonstrated by a research group of Tsinghua-Foxconn Nanotechnology Research Centre in Tsinghua University, using a Si-based semi-conducting technology compatible fabrication process.

Environmental Remediation

A CNT nano-structured sponge (nanosponge) containing sulfur and iron is more effective at soaking up water contaminants such as oil, fertilizers, pesticides and pharmaceuticals. Their magnetic properties make them easier to retrieve once the clean-up job is done. The sulfur and iron increases sponge size to around 2 centimetres (0.79 in). It also increases porosity due to beneficial defects, creating buoyancy and reusability. Iron, in the form of ferrocene makes the structure easier to control and enables recovery using magnets. Such nanosponges increase the absorption of the toxic organic solvent dichlorobenzene from water by 3.5 times. The sponges can vegetable oil up to 150 times their initial weight and can absorb engine oil as well.

Earlier, a magnetic boron-doped MWNT nanosponge that could absorb oil from water. The sponge was grown as a forest on a substrate via chemical vapor disposition. Boron puts kinks and elbows into the tubes as they grow and promotes the formation of covalent bonds. The nanosponges retain their elastic property after 10,000 compressions in the lab. The sponges are both superhydrophobic, forcing them to remain at the water's surface and oleophilic, drawing oil to them.

Other Applications

Carbon nanotubes have been implemented in nanoelectromechanical systems, including mechanical memory elements (NRAM being developed by Nantero Inc.) and nanoscale electric motors. In May 2005, Nanomix Inc. placed on the market a hydrogen sensor that integrated carbon nanotubes on a silicon platform. Since then, Nanomix has been patenting many such sensor applications, such as in the field of carbon dioxide, nitrous oxide, glucose, DNA detection, etc.

Eikos Inc of Franklin, Massachusetts and Unidym Inc. of Silicon Valley, California are developing transparent, electrically conductive films of carbon nanotubes to replace indium tin oxide (ITO). Carbon nanotube films are substantially more mechanically robust than ITO films, making them ideal for high-reliability touchscreens and flexible displays. Printable water-based inks of carbon nanotubes are desired to enable the production of these films to replace ITO. Nanotube films show promise for use in displays for computers, cell phones, PDAs, and ATMs.

A nanoradio, a radio receiver consisting of a single nanotube, was demonstrated in 2007.

A flywheel made of carbon nanotubes could be spun at extremely high velocity on a floating magnetic axis in a vacuum, and potentially store energy at a density approaching that of conventional fossil fuels. Since energy can be added to and removed from flywheels very efficiently in the form of electricity, this might offer a way of storing electricity, making the electrical grid more efficient and variable power suppliers (like wind turbines) more useful in meeting energy needs. The practicality of this depends heavily upon the cost of making massive, unbroken nanotube structures, and their failure rate under stress.

Carbon nanotube springs have the potential to indefinitely store elastic potential energy at ten times the density of lithium-ion batteries with flexible charge and discharge rates and extremely high cycling durability.

Ultra-short SWNTs (US-tubes) have been used as nanoscaled capsules for delivering MRI contrast agents in vivo.

Carbon nanotubes provide a certain potential for metal-free catalysis of inorganic and organic reactions. For instance, oxygen groups attached to the surface of carbon nanotubes have the potential to catalyze oxidative dehydrogenations or selective oxidations. Nitrogen-doped carbon nanotubes may replace platinum catalysts used to reduce oxygen in fuel cells. A forest of vertically aligned nanotubes can reduce oxygen in alkaline solution more effectively than platinum, which has been used in such applications since the 1960s. Here, the nanotubes have the added benefit of not being subject to carbon monoxide poisoning.

Wake Forest University engineers are using multiwalled carbon nanotubes to enhance the brightness of field-induced polymer electroluminescent technology, potentially offering a step forward in the search for safe, pleasing, high-efficiency lighting. In this technology, moldable polymer matrix emits light when exposed to an electrical current. It could eventually yield high-efficiency lights without the mercury vapor of compact fluorescent lamps or the bluish tint of some fluorescents and LEDs, which has been linked with circadian rhythm disruption.

Candida albicans has been used in combination with carbon nanotubes (CNT) to produce stable electrically conductive bio-nano-composite tissue materials that have been used as temperature sensing elements.

Discovery

A 2006 editorial written by Marc Monthioux and Vladimir Kuznetsov in the journal *Carbon* described the interesting and often-misstated origin of the carbon nanotube. A large percentage of academic and popular literature attributes the discovery of hollow, nanometre-size tubes composed of graphitic carbon to Sumio Iijima of NEC in 1991.

In 1952 L. V. Radushkevich and V. M. Lukyanovich published clear images of 50 nanometre diameter tubes made of carbon in the Soviet *Journal of Physical Chemistry*. This discovery was largely unnoticed, as the article was published in the Russian language, and Western scientists' access to Soviet press was limited during the Cold War. It is likely that carbon nanotubes were produced before this date, but the invention of the transmission electron microscope (TEM) allowed direct visualization of these structures.

Carbon nanotubes have been produced and observed under a variety of conditions prior to 1991. A paper by Oberlin, Endo, and Koyama published in 1976 clearly showed hollow carbon fibres with nanometre-scale diameters using a vapor-growth technique. Additionally, the authors show a TEM image of a nanotube consisting of a single wall of graphene. Later, Endo has referred to this image as a single-walled nanotube.

In 1979, John Abrahamson presented evidence of carbon nanotubes at the 14th Biennial Conference of Carbon at Pennsylvania State University. The conference paper described carbon nanotubes as carbon fibres that were produced on carbon anodes during arc discharge. A characterization of these fibres was given as well as hypotheses for their growth in a nitrogen atmosphere at low pressures.

In 1981, a group of Soviet scientists published the results of chemical and structural characterization of carbon nanoparticles produced by a thermocatalytical disproportionation of carbon monoxide. Using TEM images and XRD patterns, the authors suggested that their "carbon multi-layer tubular crystals" were formed by rolling graphene layers into cylinders. They speculated that by rolling graphene layers into a cylinder, many different arrangements of graphene hexagonal nets are possible. They suggested two possibilities of such arrangements: circular arrangement (armchair nanotube) and a spiral, helical arrangement (chiral tube).

In 1987, Howard G. Tennett of Hyperion Catalysis was issued a U.S. patent for the production of "cylindrical discrete carbon fibrils" with a "constant diameter between about 3.5 and about 70 nanometres..., length 10^2 times the diameter, and an outer region of multiple essentially continuous layers of ordered carbon atoms and a distinct inner core...."

Iijima's discovery of multi-walled carbon nanotubes in the insoluble material of arc-burned graphite rods in 1991 and Mintmire, Dunlap, and White's independent prediction that if single-walled carbon nanotubes could be made, then they would exhibit remarkable conducting properties helped create the initial buzz that is now associated with carbon nanotubes. Nanotube research accelerated greatly following the independent discoveries by Bethune at IBM and Iijima at NEC of *single-walled* carbon nanotubes and methods to specifically produce them by adding transition-metal catalysts to the carbon in an arc discharge. The arc discharge technique was well-known to produce the famed Buckminster fullerene on a preparative scale, and these results appeared to extend the run of accidental discoveries relating to fullerenes. The original observation of fullerenes in mass spectrometry was not anticipated, and the first mass-production technique by Krätschmer and Huffman was used for several years before realising that it produced fullerenes.

The discovery of nanotubes remains a contentious issue. Many believe that Iijima's report in 1991 is of particular importance because it brought carbon nanotubes into the awareness of the scientific community as a whole.

Planar Chiral Metamaterials

Planar chirality is the special case of chirality for two dimensions. Most fundamentally, planar chirality is a mathematical term, finding use in chemistry, physics and related physical sciences, for example, in astronomy, optics and metamaterials. Recent occurrences in latter two fields are dominated by microwave applications and micro- and nanostructured planar interfaces for visible and infrared light.

Planar Chirality in Chemistry

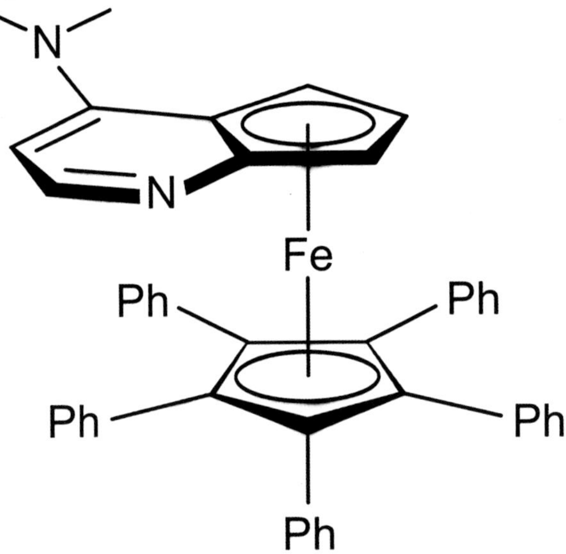

Figure: *An planar chiral derivative of ferrocene, used for kinetic resolution of some racemic secondary alcohols*

This term is used in chemistry contexts, e.g., for a chiral molecule lacking an asymmetric carbon atom, but possessing two non-coplanar rings that are each dissymmetric and which cannot easily rotate about the chemical bond connecting them: 2,2'-dimethylbiphenyl is perhaps the simplest example of this case. Planar chirality is also exhibited by molecules like (*E*)-cyclooctene, some di- or poly-substituted metallocenes, and certain monosubstituted paracyclophanes. Nature rarely provides planar chiral molecules, cavicularin being an exception.

Assigning the Configuration of Planar Chiral Molecules

To assign the configuration of a planar chiral molecule, begin by selecting the pilot atom, which is the highest priority of the atoms that is not in the plane, but are directly attached to an atom in the plane. Next, assign the priority of three adjacent in-plane atoms, starting with the atom attached to the pilot atom as priority 1, and preferentially assigning in order of highest priority if there is a choice. When viewed from the side of the pilot atom, if the three atoms form a clockwise direction when followed in order of priority, the molecule is assigned as R, otherwise it is assigned as S.

Planar Chirality in Optics and Metamaterials

The study of planar chiral metamaterials has revealed that planar chirality is associated with an optical effect: the directionally asymmetric transmission (reflection and absorption) of circularly polarized waves. Planar chiral metamaterials, which are also anisotropic and lossy exhibit different total transmission (reflection and absorption) levels for the same circularly polarized wave incident on their front and back. The asymmetric transmission phenomenon arises from different, e.g. left-to-right, circular polarization conversion efficiencies for opposite propagation directions of the incident wave and therefore the effect is referred to as circular conversion dichroism. Like the twist of a planar chiral pattern appears reversed for opposite directions of observation, planar chiral metamaterials have interchanged properties for left-handed and right-handed circularly polarized waves that are incident on their front and back. In particular left-handed and right-handed circularly polarized waves experience opposite directional transmission (reflection and absorption) asymmetries .

The Surface Plasmon

Surface plasmons (SPs) are coherent electron oscillations that exist at the interface between any two materials where the real part of the dielectric function changes sign across the interface (e.g. a metal-dielectric interface, such as a metal sheet in air). SPs have lower energy than bulk (or volume) plasmons which quantise the longitudinal electron oscillations about positive ion cores within the bulk of an electron gas (or plasma). The charge motion in a surface plasmon always creates electromagnetic fields outside (as well as inside) the metal. The *total* excitation, including both the charge motion and associated electromagnetic field, is called either a surface plasmon polariton at a planar interface, or a localized surface plasmon for the closed surface of a small particle.

The existence of surface plasmons was first predicted in 1957 by Rufus Ritchie. In the following two decades, surface plasmons were extensively studied by many scientists, the foremost of whom were T. Turbadar in the 1950s and 1960s, and Heinz Raether, E. Kretschmann, and A. Otto in the 1960s and 1970s. Information transfer in nanoscale structures, similar to photonics, by means of surface plasmons, is referred to as plasmonics.

Near-Field Scanning Optical Microscope

Near-field scanning optical microscopy (NSOM/SNOM) is a microscopy technique for nanostructure investigation that breaks the far field resolution limit by exploiting the properties of evanescent waves. This is done by placing the detector very close (distance much smaller than wavelength λ) to the specimen surface. This allows for the surface inspection with high spatial, spectral and temporal resolving power. With this technique, the resolution of the image is limited by the size of the detector aperture and not by the wavelength of the illuminating light. In particular, lateral resolution of 20 nm and vertical resolution of 2–5 nm have been demonstrated. As in optical microscopy, the contrast mechanism can be easily adapted to study different properties, such as refractive index, chemical structure and local stress. Dynamic properties can also be studied at a sub-wavelength scale using this technique.

Edward Hutchinson Synge, a scientist, is given credit for conceiving and developing the idea for an imaging instrument that would image by exciting and collecting diffraction in the near field. His original idea, proposed in 1928, was based upon the usage of intense nearly planar light from an arc under pressure behind a thin, opaque metal film with a small orifice of about 100 nm. The orifice was to remain within 100 nm of the surface, and information was to be collected by point-by-point scanning. He foresaw the illumination and the detector movement being the biggest technical difficulties. John A. O'Keefe also developed similar theories in 1956. He thought the moving of the pinhole or the detector when it is so close to the sample would be the most likely issue that could prevent the realisation of such an instrument. It was Ash and Nicholls who, in 1972, first broke the Abbe's diffraction limit using radiation with wavelength of 3 cm. A line grating was resolved with a resolution of $\lambda_0/60$. A decade later, a patent on an *optical* near-field microscope was filed by Pohl, followed in 1984 by the first paper that used visible radiation for near field scanning. The near-field optical (NFO) microscope involved a subwavelength aperture at the apex of a metal coated sharply pointed transparent tip, and a feedback mechanism to maintain a constant distance of a few nanometres between the sample and the probe. Lewis et al. were also aware of the potential of an NFO microscope at this time. They reported first results in 1986 confirming super-resolution. In both experiments, details below 50 nm (about $\lambda_0/10$) in size could be recognised.

Theory

According to Abbe's theory of image formation, developed in 1873, the resolving capability of an optical component is ultimately limited by the spreading out of each image point due to diffraction. Unless the aperture of the optical component is large enough to collect all the diffracted light, the finer aspects of the image will not correspond exactly to the object. The minimum resolution (d) for the optical component are thus limited by its aperture size, and expressed by the Rayleigh criterion:

$$d = 0.61\frac{\lambda_0}{NA}$$

Here, λ_0 is the wavelength in vacuum; NA is the numerical aperture for the optical component (maximum 1.3–1.4 for modern objectives with a very high magnification factor). Thus, the resolution limit is usually around $\lambda_0/2$ for conventional optical microscopy.

This treatment only assumes the light diffracted into the far-field that propagates without any restrictions. NSOM makes use of evanescent or non propagating fields that exist only near the surface of the object. These fields carry the high frequency spatial information about the object and have intensities that drop off exponentially with distance from the object.

Because of this, the detector must be placed very close to the sample in the near field zone, typically a few nanometres. As a result, near field microscopy remains primarily a surface inspection technique. The detector is then rastered across the sample using a piezoelectric stage. The scanning can either be done at a constant height or with regulated height by using a feedback mechanism.

Modes of Operation

Aperture and Apertureless Operation

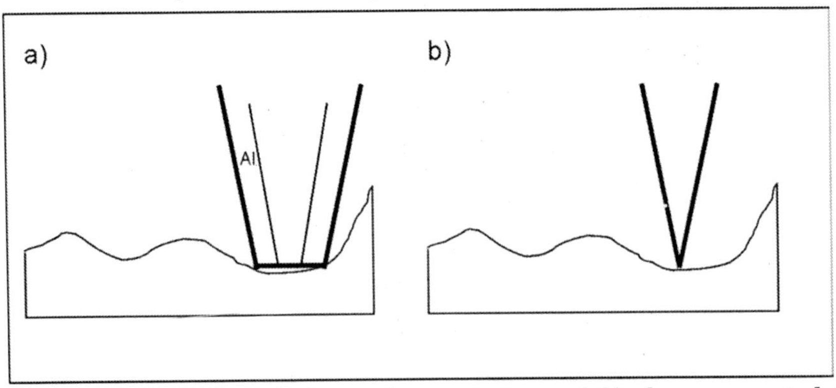

Figure: *Sketch of a) typical metal-coated tip, and b) sharp uncoated tip.*

There exist NSOM which can be operated in so-called aperture mode and NSOM for operation in a non-aperture mode. As illustrated, the tips used in the apertureless mode are very sharp and do not have a metal coating.

Though there are many issues associated with the apertured tips (heating, artifacts, contrast, sensitivity, topology and interference amongst others), aperture mode remains more popular. This is primarily because apertureless mode is even more complex to set up and operate, and is not understood as well. There are five primary modes of apertured NSOM operation and four primary modes of apertureless NSOM operation. The major ones are illustrated in the next figure.

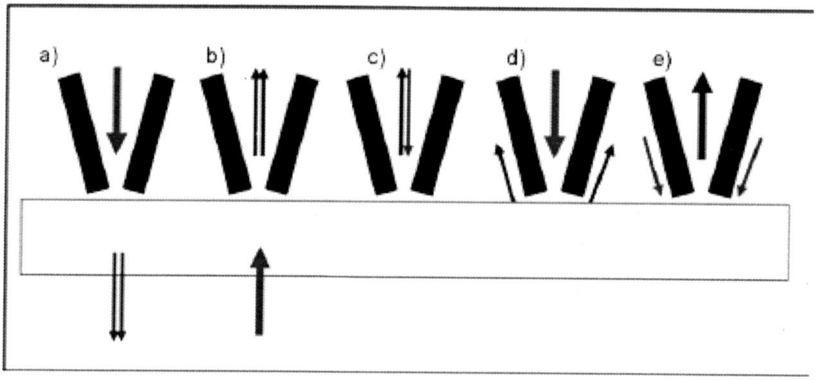

Figure: *Apertured modes of operation: a) illumination, b) collection, c) illumination collection, d) reflection and e) reflection collection.*

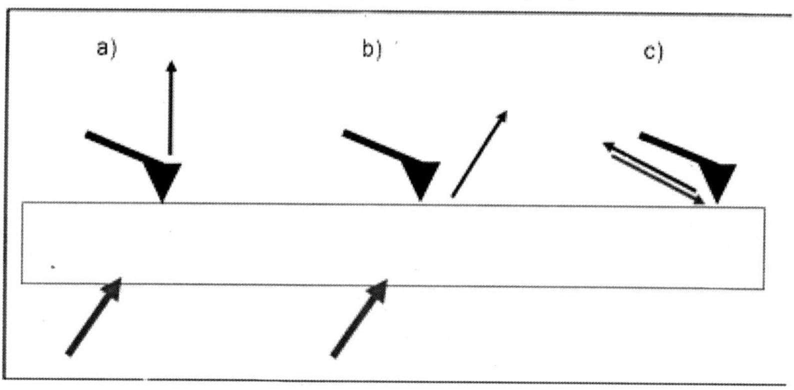

Figure: *Apertureless modes of operation: a) photon tunneling (PSTM) by a sharp transparent tip, b) PSTM by sharp opaque tip on smooth surface, and c) scanning interferometric apertureless microscopy with double modulation.*

Other types of NSOM operation utilise "active tip" schemes, where the tip is functionalized with active light sources such as a fluorescent dye or even a light emitting diode that enables fluorescence excitation.

Feedback Mechanisms

Feedback mechanisms are usually used to achieve high resolution and artifact free images since the detector must be positioned within a few nanometres of the surfaces. Some of these mechanisms are:

- Constant force feedback: This mode is very similar to the feedback mechanism used in atomic force microscopy (AFM). Experiments can be performed in contact, intermittent contact, and non-contact modes.
- Shear force feedback: In this mode, a tuning fork is mounted alongside the tip and made to oscillate at its resonance frequency. The amplitude is closely related to the tip-surface distance, and thus used as a feedback mechanism.

Contrast

It is possible to take advantage of the various contrast techniques available to optical microscopy through NSOM but with much higher resolution. By using the change in the polarization of light or the intensity of the light as a function of the incident wavelength, it is possible to make use of contrast enhancing techniques such as staining, fluorescence, phase contrast and differential interference contrast. It is also possible to provide contrast using the change in refractive index, reflectivity, local stress and magnetic properties amongst others.

Instrumentation and Standard Setup

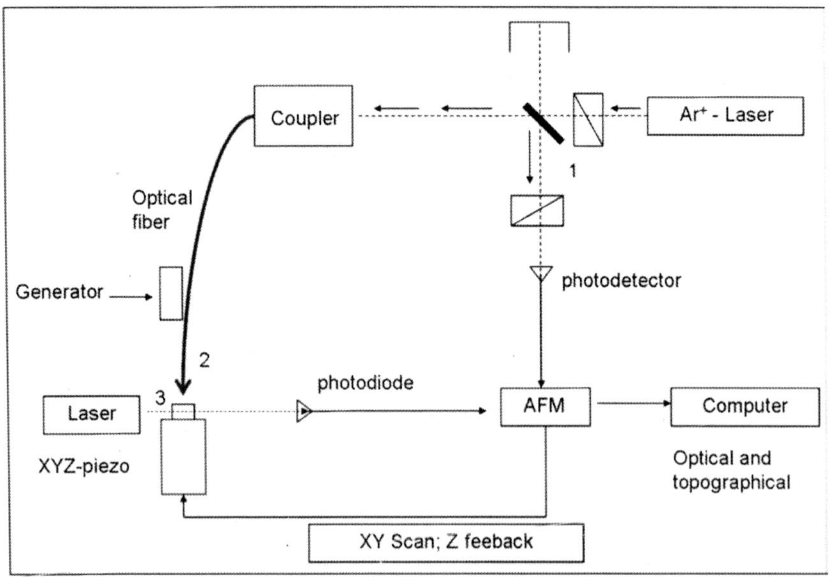

***Figure:** Block diagram of an apertureless reflection-back-to-the-fibre NSOM setup with shear-force distance control and cross-polarization; 1: beam splitter and crossed polarizers; 2: shear-force arrangement; 3: sample mount on a piezo stage.*

The primary components of an NSOM setup are the light source, feedback mechanism, the scanning tip, the detector and the piezoelectric sample stage. The light source is usually a laser focused into an optical fibre through a polarizer, a beam splitter and a coupler. The polarizer and the beam splitter would serve to remove stray light from the returning reflected light. The scanning tip, depending upon the operation mode, is usually a pulled or stretched optical fibre coated with metal except at the tip or just a standard AFM cantilever with a hole in the centre of the pyramidal tip. Standard optical detectors, such as avalanche photodiode, photomultiplier tube (PMT) or CCD, can be used. Highly specialised NSOM techniques, Raman NSOM for example, have much more stringent detector requirements.

Near-Field Spectroscopy

As the name implies, information is collected by spectroscopic means instead of imaging in the near field regime. Through Near Field Spectroscopy (NFS), one can probe spectroscopically with subwavelength resolution. Raman SNOM and fluorescence SNOM are two of the most popular NFS techniques as they allow for the identification of nanosized features with chemical contrast. Some of the common near field spectroscopic techniques are:

- *Direct local Raman NSOM:* Aperture Raman NSOM is limited by very hot and blunt tips, and by long collection times. However, apertureless NSOM can be used to achieve high Raman scattering efficiency factors (around 40). Topological artifacts make it hard to implement this technique for rough surfaces.
- *Surface enhanced Raman spectroscopy (SERS) NSOM:* This technique can be used in an apertureless shear-force NSOM setup, or by using an AFM tip coated with gold. The Raman signal is found to be significantly enhanced under the AFM tip. This technique has been used to give local variations in the Raman spectra under a single-walled nanotube. A highly sensitive optoacoustic spectrometre must be used for the detection of the Raman signal.
- *Fluorescence NSOM:* This highly popular and sensitive technique makes use of the fluorescence for near field imaging, and is especially suited for biological applications. The technique of choice here is the apertureless back to the fibre emission in constant shear force mode. This technique uses merocyanine based dyes embedded in an appropriate resin. Edge filters are used for removal of all primary laser light. Resolution as low as 10 nm can be achieved using this technique.
- *Near field infrared spectrometry and near field dielectric microscopy*

Nanomagnetism

A nanomagnet is a submicrometric system that presents spontaneous magnetic order (magnetization) at zero applied magnetic field (remanence). The small size of nanomagnets prevents the formation of magnetic domains. The magnetization dynamics of sufficiently small nanomagnets at low temperatures, typically single-molecule magnets, presents quantum phenomena, such as macroscopic spin tunnelling. At larger temperatures, the magnetization undergoes random thermal fluctuations (superparamagnetism) which present a limit for the use of nanomagnets for permanent information storage.

Canonical examples of nanomagnets are grains of ferromagnetic metals (iron, cobalt, and nickel) and single-molecule magnets. The vast majority of nanomagnets feature transition metal (titanium, vanadium, chromium, manganese, iron, cobalt or nickel) or rare earth (Gd, Eu, Er) magnetic atoms.

Attending to the number of magnetic atoms, the smallest nanomagnets reported so far are double decker phthalocyanes with only one rare earth atom. Canonical single-molecule magnets are the so-called Mn12 and Fe8 systems, with 12 and 8 transition metal atoms each and both with spin 10 (S=10) ground states.

The phenomenon of zero field magnetization requires three conditions:

- (1) A ground state with finite spin
- (2) a magnetic anisotropy energy barrier
- (3) long spin relaxation time.

Conditions (1) and (2), but not (3), have been demonstrated in a number of nanostructures, such as nanoparticles, nanoislands, and quantum dots with a controlled number of magnetic atoms (between 1 and 10). A nanomagnet can have enhanced electronic properties due to size effect, such as long spin relaxation time of conduction electron, which may be useful for nano-scale spintronic device.

Magnetic Nanostructures

An important focus of the group focus lies in the investigation of the electronic and magnetic properties of nanostructures at surfaces. Using state-of-the-art commercial and home-made density functional theory and kinetic Monte Carlo codes (software) we study, among other things, electronic properties of nanoscale surface structures, and spin dependent effects therein, to better understand how they can be controlled, measured, and possibly applied in modern technological challenges, such as the advancement of spin electronics (spintronics) and novel data storage development. And since for success of spintronics it is not sufficient to understand the intrinsic magnetic and electronic properties of nanostructures we also strive to find adequate means of controlling those properties externally.

Spin Dependent Effects

To understand and control magnetism of surface nanostructures one has to understand and control the electrons at the surface. Numerous interesting effects can be observed and put to use at surfaces that contain the so-called Shockley surface state - a state localized near the surface and behaving as a quasi-2-dimensional electron gas. The electrons in this state scatter at surface impurities and if the impurities form a closed geometry, the surface state electrons can become confined to that geometry. Carefully controlling the confinement shape one can deliberately channel and focus electrons. For example, it has been shown experimentally, that an elliptic atomic (quantum) corral can coherently transfer the information about the magnetic (Kondo) state of a single atom residing in one of its focal points to the other focal point. Using the Korringa-Kohn-Rostoker Green's function method we can fully *ab initio* calculate such complex multiple scattering problems.

This amazing property of quantum corrals can be used to tailor magnetic interaction between single adatoms if those are placed in the focuses of the corral or even a part of a corral. Our calculations show that by carefully adjusting the size and shape of the corral one can enhance, quench or even reverse the coupling between individual spins at the surface. Yet it is not always necessary to use corrals, that are extremely hard to produce experimentally, to confine surface electrons. Natural or self-assembled surface structures can serve the same purpose. So, for example, our calculations have proven, that quantum confinement of surface state electrons on hexagonal copper islands can be used to tailor the interaction between single magnetic adatoms adsorbed on top of them.

While confinement of surface state electrons by paramagnetic structures can aid in controlling the interaction of single magnetic impurities adsorbed within, the confinement itself can also be spin-dependent. So, f. e., our calculations have predicted that the spin-dependent confinement on triangular cobalt islands leads to a strong variation of the local magnetization within a single island/nanostructure. Quite recently our predictions have been confirmed by our experimental colleagues.

7

Nanocomposites and Fibres

A nanocomposite is a multiphase solid material where one of the phases has one, two or three dimensions of less than 100 nanometres (nm), or structures having nano-scale repeat distances between the different phases that make up the material. In the broadest sense this definition can include porous media, colloids, gels and copolymers, but is more usually taken to mean the solid combination of a bulk matrix and nano-dimensional phase(s) differing in properties due to dissimilarities in structure and chemistry. The mechanical, electrical, thermal, optical, electrochemical, catalytic properties of the nanocomposite will differ markedly from that of the component materials. Size limits for these effects have been proposed, <5 nm for catalytic activity, <20 nm for making a hard magnetic material soft, <50 nm for refractive index changes, and <100 nm for achieving superparamagnetism, mechanical strengthening or restricting matrix dislocation movement.

Nanocomposites are found in nature, for example in the structure of the abalone shell and bone. The use of nanoparticle-rich materials long predates the understanding of the physical and chemical nature of these materials. Jose-Yacaman *et al.* investigated the origin of the depth of colour and the resistance to acids and bio-corrosion of Maya blue paint, attributing it to a nanoparticle mechanism. From the mid-1950s nanoscale organo-clays have been used to control flow of polymer solutions (e.g. as paint viscosifiers) or the constitution of gels (e.g. as a thickening substance in cosmetics, keeping the preparations in homogeneous form). By the 1970s polymer/clay composites were the topic of textbooks, although the term "nanocomposites" was not in common use. In mechanical terms, nanocomposites differ from conventional composite materials due to the exceptionally high surface to volume ratio of the reinforcing phase and/or its exceptionally high aspect ratio. The reinforcing material can be made up of particles (e.g. minerals), sheets (e.g. exfoliated clay stacks) or fibres (e.g. carbon nanotubes or electrospun fibres). The area of the interface between the matrix and reinforcement phase(s) is typically an order of magnitude greater than for conventional composite materials. The matrix material properties are significantly affected in the vicinity of the reinforcement. Ajayan *et al.* note that with polymer nanocomposites, properties related to local chemistry, degree of thermoset cure, polymer chain mobility, polymer chain conformation, degree of polymer chain ordering or crystallinity can all vary significantly and continuously from the interface with the reinforcement into the bulk of the matrix.

This large amount of reinforcement surface area means that a relatively small amount of nanoscale reinforcement can have an observable effect on the macroscale properties of the composite. For example, adding carbon nanotubes improves the electrical and thermal conductivity. Other kinds of nanoparticulates may result in enhanced optical properties, dielectric properties, heat resistance or mechanical properties such as stiffness, strength and resistance to wear and damage. In general, the nano reinforcement is dispersed into the matrix during processing. The percentage by weight (called *mass fraction*) of the nanoparticulates introduced can remain very low (on the order of 0.5% to 5%) due to the low filler percolation threshold, especially for the most commonly used non-spherical, high aspect ratio fillers (e.g. nanometre-thin platelets, such as clays, or nanometre-diameter cylinders, such as carbon nanotubes).

Ceramic-Matrix Nanocomposites

In this group of composites the main part of the volume is occupied by a ceramic, i.e. a chemical compound from the group of oxides, nitrides, borides, silicides etc.. In most cases, ceramic-matrix nanocomposites encompass a metal as the second component. Ideally both components, the metallic one and the ceramic one, are finely dispersed in each other in order to elicit the particular nanoscopic properties. Nanocomposite from these combinations were demonstrated in improving their optical, electrical and magnetic properties as well as tribological, corrosion-resistance and other protective properties.

The binary phase diagram of the mixture should be considered in designing ceramic-metal nanocomposites and measures have to be taken to avoid a chemical reaction between both components. The last point mainly is of importance for the metallic component that may easily react with the ceramic and thereby lose its metallic character. This is not an easily obeyed constraint, because the preparation of the ceramic component generally requires high process temperatures. The most safe measure thus is to carefully choose immiscible metal and ceramic phases. A good example for such a combination is represented by the ceramic-metal composite of TiO_2 and Cu, the mixtures of which were found immiscible over large areas in the Gibbs' triangle of Cu-O-Ti.

The concept of ceramic-matrix nanocomposites was also applied to thin films that are solid layers of a few nm to some tens of μm thickness deposited upon an underlying substrate and that play an important role in the functionalization of technical surfaces. Gas flow sputtering by the hollow cathode technique turned out as a rather effective technique for the preparation of nanocomposite layers. The process operates as a vacuum-based deposition technique and is associated with high deposition rates up to some μm/s and the growth of nanoparticles in the gas phase. Nanocomposite layers in the ceramics range of composition were prepared from TiO_2 and Cu by the hollow cathode technique that showed a high mechanical hardness, small coefficients of friction and a high resistance to corrosion.

Metal-Matrix Nanocomposites

Metal matrix nanocomposites can also be defined as reinforced metal matrix composites. This type of composites can be classified as continuous and non-continuous reinforced materials. One of the more important nanocomposites is Carbon nanotube metal matrix composites, which is an emerging new material that is being developed to take advantage of the high tensile strength and electrical conductivity of carbon nanotube materials. Critical to the realisation of CNT-MMC possessing optimal properties in these areas are the development of synthetic techniques that are (a) economically producible, (b) provide for a homogeneous dispersion of nanotubes in the metallic matrix, and (c) lead to strong interfacial adhesion between the metallic matrix and the carbon nanotubes. In addition to carbon nanotube metal matrix composites, boron nitride reinforced metal matrix composites and carbon nitride metal matrix composites are the new research areas on metal matrix nanocomposites.

A recent study, comparing the mechanical properties (Young's modulus, compressive yield strength, flexural modulus and flexural yield strength) of single- and multi-walled reinforced polymeric (polypropylene fumarate—PPF) nanocomposites to tungsten disulfide nanotubes reinforced PPF nanocomposites suggest that tungsten disulfide nanotubes reinforced PPF nanocomposites possess significantly higher mechanical properties and tungsten disulfide nanotubes are better reinforcing agents than carbon nanotubes. Increases in the mechanical properties can be attributed to a uniform dispersion of inorganic nanotubes in the polymer matrix (compared to carbon nanotubes that exist as micron sized aggregates) and increased crosslinking density of the polymer in the presence of tungsten disulfide nanotubes (increase in crosslinking density leads to an increase in the mechanical properties). These results suggest that inorganic nanomaterials, in general, may be better reinforcing agents compared to carbon nanotubes.

Another kind of nanocomposite is the energetic nanocomposite, generally as a hybrid sol–gel with a silica base, which, when combined with metal oxides and nano-scale aluminum powder, can form *superthermite* materials.

Polymer-Matrix Nanocomposites

In the simplest case, appropriately adding nanoparticulates to a polymer matrix can enhance its performance, often dramatically, by simply capitalizing on the nature and properties of the nanoscale filler (these materials are better described by the term *nanofilled polymer composites*). This strategy is particularly effective in yielding high performance composites, when good dispersion of the filler is achieved and the properties of the nanoscale filler are substantially different or better than those of the matrix. An example of this would be reinforcing a polymer matrix by much stiffer nanoparticles of ceramics, clays, or carbon nanotubes. It should be noted that the improvement in mechanical properties may not be limited to stiffness or strength. Time-dependent properties could be improved by addition of the nanofillers. Alternatively, the enhanced crystallization behaviour under flow conditions and other physical properties of high performance nanocomposites may be mainly due to the high aspect ratio and/or the high surface area of the fillers, since nanoparticulates have extremely high surface area to volume ratios when good dispersion is achieved. Nanoparticle dispersibility in the polymer matrix is a key issue, which limits the applicable particle volume fraction and there for also the multi-functionality of the composite material. Recent research on thin films (thickness <50 micrometre) made of polymer nanocomposites has resulted in a new and scalable synthesis technique, which allows the facile incorporation of greater nano-material quantities. Such advances will enable the future development of multi-functional small scale devices (i.e. sensor, actuator, medical equipment), which rely on polymer nanocomposites. Nanoscale dispersion of filler or controlled nanostructures in the composite can introduce new physical properties and novel behaviours that are absent in the unfilled matrices. This effectively changes the nature of the original matrix (such composite materials can be better described by the term *genuine nanocomposites* or *hybrids*). Some examples of such new properties are fire resistance or flame retardancy, and accelerated biodegradability.

In a recent study, polymeric nanocomposites were fabricated using various one-dimensional carbon nanostructures such as single- and multi- walled carbon nanotubes, as well as two-dimensional carbon and inorganic nanomaterials such as graphene platelets, graphene nanoribbon, single- and multi- walled graphene oxide nanoribbons, graphene oxide nanoplatelets and molybdenum disulfide nanoplatelets as reinforcing agents, to improve the mechanical properties of poly (propylene fumarate) nanocomposites, for bone tissue engineering applications. Significant mechanical reinforcement (i.e. increases in the Young's modulus, compressive yield strength, flexural modulus and flexural yield strength) were observed at low loading concentrations (0.01-0.2 wt%) of nanomaterials. The results suggest that mechanical reinforcement is dependent on the nanostructure morphology, defects, dispersion of nanomaterials in the polymer matrix, and the cross-linking density of the polymer. In general, two-dimensional nanostructures can reinforce the polymer better than one-dimensional nanostructures, and inorganic nanomaterials are better reinforcing agents than carbon based nanomaterials. In addition to mechanical properties, multi-walled carbon nanotubes based polymer nanocomposites have also been used for the enhancement of the electrical conductivity.

Carbon Fibres and Nanotubes

Carbon fibre, alternatively graphite fibre, carbon graphite or CF, is a material consisting of fibres about 5–10 μm in diameter and composed mostly of carbon atoms. To produce carbon fibre, the carbon atoms are bonded together in crystals that are more or less aligned parallel to the long axis of the fibre as the crystal alignment gives the fibre high strength-to-volume ratio (making it strong for its size). Several thousand carbon fibres are bundled together to form a tow, which may be used by itself or woven into a fabric.

The properties of carbon fibres, such as high stiffness, high tensile strength, low weight, high chemical resistance, high temperature tolerance and low thermal expansion, make them very popular in aerospace, civil engineering, military, and motorsports, along with other competition sports. However, they are relatively expensive when compared to similar fibres, such as glass fibres or plastic fibres.

Carbon fibres are usually combined with other materials to form a composite. When combined with a plastic resin and wound or molded it forms carbon fibre reinforced polymer (often referred to as carbon fibre) which has a very high strength-to-weight ratio, and is extremely rigid although somewhat brittle. However, carbon fibres are also composed with other materials, such as with graphite to form carbon-carbon composites, which have a very high heat tolerance.

In 1958, Roger Bacon created high-performance carbon fibres at the Union Carbide Parma Technical Centre, now GrafTech International Holdings, Inc., located outside of Cleveland, Ohio. Those fibres were manufactured by heating strands of rayon until they carbonized. This process proved to be inefficient, as the resulting fibres contained only about 20% carbon and had low strength and stiffness properties. In the early 1960s, a process was developed by Dr. Akio Shindo at Agency of Industrial Science and Technology of Japan, using polyacrylonitrile (PAN) as a raw material. This had produced a carbon fibre that contained about 55% carbon.

The high potential strength of carbon fibre was realised in 1963 in a process developed by W. Watt, L. N. Phillips, and W. Johnson at the Royal Aircraft Establishment at Farnborough, Hampshire. The process was patented by the UK Ministry of Defence then licensed by the National Research Development Corporation (NRDC) to three British companies: Rolls-Royce, already making carbon fibre; Morganite; and Courtaulds. They were able to establish industrial carbon fibre production facilities within a few years, and Rolls-Royce took advantage of the new material's properties to break into the American market with its RB-211 aero-engine.

Public concern arose over the ability of British industry to make the best of this breakthrough. In 1969 a House of Commons select committee inquiry into carbon fibre prophetically asked: "How then is the nation to reap the maximum benefit without it becoming yet another British invention to be exploited more successfully overseas?" Ultimately, this concern was justified. One by one the licensees pulled out of carbon-fibre manufacture. Rolls-Royce's interest was in state-of-the-art aero-engine applications. Its own production process was to enable it to be leader in the use of carbon-fibre reinforced plastics. In-house production would typically cease once reliable commercial sources became available.

Unfortunately, Rolls-Royce pushed the state-of-the-art too far, too quickly, in using carbon fibre in the engine's compressor blades, which proved vulnerable to damage from bird impact. What seemed a great British technological triumph in 1968 quickly became a disaster as Rolls-Royce's ambitious schedule for the RB-211 was endangered. Indeed, Rolls-Royce's problems became so great that the company was eventually nationalized by the British government in 1971 and the carbon-fibre production plant was sold off to form "Bristol Composites".

Given the limited market for a very expensive product of variable quality, Morganite also decided that carbon-fibre production was peripheral to its core business, leaving Courtaulds as the only big UK manufacturer. Continuing collaboration with the staff at Farnborough proved helpful in the quest for higher quality and improvements in the speed of production as Courtaulds developed two main markets: aerospace and sports equipment.. However Courtaulds's big advantage as manufacturer of the "Courtelle" precursor now became a weakness. Courtelle's low cost and ready availability were potential advantages, but the water-based inorganic process used to produce it made the product susceptible to impurities that did not affect the organic process used by other carbon-fibre manufacturers.

Nevertheless, during the 1980s Courtaulds continued to be a major supplier of carbon fibre for the sports-goods market, with Mitsubishi its main customer until a move to expand, including building a production plant in California, turned out badly. The investment did not generate the anticipated returns, leading to a decision to pull out of the area and Courtaulds ceased carbon-fibre production in 1991. Ironically the one surviving UK carbon-fibre manufacturer continued to thrive making fibre based on Courtaulds's precursor. Inverness-based RK Carbon Fibres Ltd concentrated on producing carbon fibre for industrial applications, removing the need to compete at the quality levels reached by overseas manufacturers.

During the 1970s, experimental work to find alternative raw materials led to the introduction of carbon fibres made from a petroleum pitch derived from oil processing. These fibres contained about 85% carbon and had excellent flexural strength. Also, during this period, the Japanese Government heavily supporting carbon fibre development at home and several Japanese companies such as Toray, Nippon Carbon, Toho Rayon and Mitsubishi started their own development and production. As they subsequently advanced to become market leaders, companies in USA and Europe were encouraged to take up these activities as well, either through their own developments or contractual acquisition of carbon fibre knowledge. These companies included Hercules, BASF and Celanese USA and Akzo in Europe.

During this period of advancement, further types of carbon fibre yarn entered the global market, offering higher tensile strength and higher elastic modulus. For example T400 from Toray with a tensile strength of 4,000 MPa and M40, a modulus of 400 GPa. Intermediate carbon fibres, such as IM 600 from Toho Reyon with up to 6,000 MPa were developed. Carbon fibres from Toray, Celanese and Akzo found their way to aerospace application from secondary to primary parts first in military and later in civil aircraft as in McDonnell Douglas, Boeing and Airbus planes. By 2000 the industrial applications for high sophisticated machine parts in middle Europe was becoming more important.

Structure and Properties

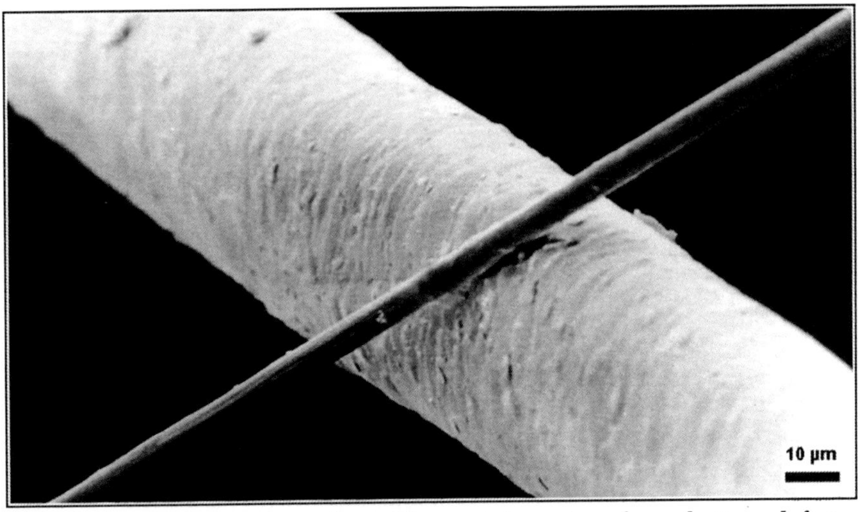

Figure: *A 6 μm diameter carbon filament (running from bottom left to top right) compared to a human hair.*

Each carbon filament thread is a bundle of many thousand carbon filaments. A single such filament is a thin tube with a diameter of 5–8 micrometres and consists almost exclusively of carbon. The earliest generation (e.g. T300, HTA and AS4) had diameters of 16–22 micrometres. Later fibres (e.g. IM6 or IM600) have diameters that are approximately 5 micrometres.

The atomic structure of carbon fibre is similar to that of graphite, consisting of sheets of carbon atoms (graphene sheets) arranged in a regular hexagonal pattern, the difference being in the way these sheets interlock. Graphite is a crystalline material in which the sheets are stacked parallel to one another in regular fashion. The intermolecular forces between the sheets are relatively weak Van der Waals forces, giving graphite its soft and brittle characteristics.

Depending upon the precursor to make the fibre, carbon fibre may be turbostratic or graphitic, or have a hybrid structure with both graphitic and turbostratic parts present. In turbostratic carbon fibre the sheets of carbon atoms are haphazardly folded, or crumpled, together. Carbon fibres derived from Polyacrylonitrile (PAN) are turbostratic, whereas carbon fibres derived from mesophase pitch are graphitic after heat treatment at temperatures exceeding 2200 °C. Turbostratic carbon fibres tend to have high tensile strength, whereas heat-treated mesophase-pitch-derived carbon fibres have high Young's modulus (i.e., high stiffness or resistance to extension under load) and high thermal conductivity.

Applications

Figure: *Tail of an RC helicopter, made of Carbon fibre reinforced polymer*

The global demand on carbon fibre composites was valued at roughly US$10.8 billion in 2009, which declined 8–10% from the previous year. It is expected to reach US$13.2 billion by 2012 and to increase to US$18.6 billion by 2015 with an annual growth rate of 7% or more. Strongest demands come from aircraft & aerospace, wind energy, as well as from the automotive industry with optimized resin systems.

Composite Materials

Carbon fibre is most notably used to reinforce composite materials, particularly the class of materials known as carbon fibre or graphite reinforced polymers. Non-polymer materials can also be used as the matrix for carbon fibres.

Due to the formation of metal carbides and corrosion considerations, carbon has seen limited success in metal matrix composite applications. Reinforced carbon-carbon (RCC) consists of carbon fibre-reinforced graphite, and is used structurally in high-temperature applications.

The fibre also finds use in filtration of high-temperature gases, as an electrode with high surface area and impeccable corrosion resistance, and as an anti-static component. Molding a thin layer of carbon fibres significantly improves fire resistance of polymers or thermoset composites because a dense, compact layer of carbon fibres efficiently reflects heat.

The increasing use of carbon fibre composites is displacing aluminum from aerospace applications in favour of other metals because of galvanic corrosion issues.

Textiles

Precursors for carbon fibres are polyacrylonitrile (PAN), rayon and pitch. Carbon fibre filament yarns are used in several processing techniques: the direct uses are for prepregging, filament winding, pultrusion, weaving, braiding, etc. Carbon fibre yarn is rated by the linear density (weight per unit length, i.e. 1 g/1000 m = 1 tex) or by number of filaments per yarn count, in thousands.

For example, 200 tex for 3,000 filaments of carbon fibre is three times as strong as 1,000 carbon filament yarn, but is also three times as heavy. This thread can then be used to weave a carbon fibre filament fabric or cloth. The appearance of this fabric generally depends on the linear density of the yarn and the weave chosen. Some commonly used types of weave are twill, satin and plain. Carbon filament yarns can be also knitted or braided.

Microelectrodes

Carbon fibres are used for fabrication of carbon-fibre microelectrodes. In this application typically a single carbon fibre with diameter of 5–7 μm is sealed in a glass capillary. At the tip the capillary is either sealed with epoxy and polished to make carbon-fibre disk microelectrode or the fibre is cut to a length of 75–150 μm to make carbon-fibre cylinder electrode. Carbon-fibre microelectrodes are used either in amperometry or fast-scan cyclic voltammetry for detection of biochemical signalling.

Catalysis

PAN-based nanofibres can efficiently catalyze the first step in the making of synthetic gasoline (syngas) and other energy-rich products out of carbon dioxide. The process uses a "co-catalyst" system in three steps: (1) EMIM–CO2 complex formation; (2) adsorption of EMIM–CO2 complex on reduced carbon atoms and (3) Carbon monoxide formation.

The first step uses an ionic liquid, while graphitic carbon structures doped with other reactive atoms replaced silver to produce the final output. The carbon nanofibre catalyst exhibited negligible overpotential (0.17 V) for carbon dioxide reduction and more than an order of magnitude higher current density compared with silver under similar experimental conditions. The reduction derived from the reduced carbons rather than to electronegative nitrogen dopants. The performance came from the nanofibrillar structure and high binding energy of key intermediates to the carbon nanofibre surfaces.

Synthesis

Each carbon filament is produced from a precursor polymer such as polyacrylonitrile (PAN), rayon, or petroleum pitch. For synthetic polymers such as PAN or rayon, the precursor is first spun into filament yarns, using chemical and mechanical processes to initially align the polymer atoms in a way to enhance the final physical properties of the completed carbon fibre. Precursor compositions and mechanical processes used during spinning filament yarns may vary among manufacturers. After drawing or spinning, the polymer filament yarns are then heated to drive off non-carbon atoms (carbonization), producing the final carbon fibre. The carbon fibres filament yarns may be further treated to improve handling qualities, then wound on to bobbins.

A common method of manufacture involves heating the spun PAN filaments to approximately 300 °C in air, which breaks many of the hydrogen bonds and oxidizes the material. The oxidized PAN is then placed into a furnace having an inert atmosphere of a gas such as argon, and heated to approximately 2000 °C, which induces graphitization of the material, changing the molecular bond structure. When heated in the correct conditions, these chains bond side-to-side (ladder polymers), forming narrow graphene sheets which eventually merge to form a single, columnar filament. The result is usually 93–95% carbon. Lower-quality fibre can be manufactured using pitch or rayon as the precursor instead of PAN. The carbon can become further enhanced, as high modulus, or high strength carbon, by heat treatment processes. Carbon heated in the range of 1500–2000 °C (carbonization) exhibits the highest tensile strength (820,000 psi, 5,650 MPa or N/mm^2), while carbon fibre heated from 2500 to 3000 °C (graphitizing) exhibits a higher modulus of elasticity (77,000,000 psi or 531 GPa or 531 kN/mm²).

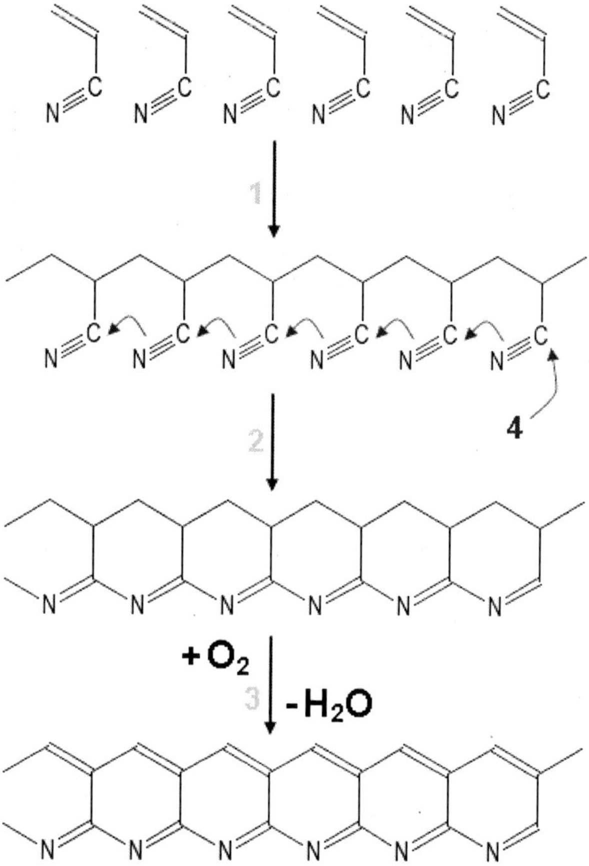

***Figure:** Synthesis of carbon fibre from polyacrylonitrile (PAN): 1) Polymerization of acrylonitrile to PAN, 2) Cyclization during low temperature process, 3) High temperature oxidative treatment of carbonization (hydrogen is removed). After this, process of graphitization starts where nitrogen is removed and chains are joined into graphite planes.*

Manufacturers

Major manufacturers of carbon fibres include Cytec Industries, Formosa Plastics, Hexcel, Mitsubishi Rayon, SGL Carbon, Toho Tenax, Toray Industries and Zoltek. Manufacturers typically make different grades of fibres for different applications. Higher modulus carbon fibres are typically more expensive.

Clay Nanocomposite Materials

Nanoclays are nanoparticles of layered mineral silicates. Depending on chemical composition and nanoparticle morphology, nanoclays are organised into several classes such as montmorillonite, bentonite, kaolinite, hectorite, and halloysite. Organically-modified nanoclays (organoclays) are an attractive class of hybrid organic-inorganic nanomaterials with potential uses in polymer nanocomposites, as rheological modifiers, gas absorbents and drug delivery carriers.

Montmorillonite Nanoclays

Plate-like montmorillonite is the most common nanoclay used in materials applications. Montmorillonite consists of ~ 1 nm thick aluminosilicate layers surface-substituted with metal cations and stacked in ~ 10 μm-sized multilayer stacks. Depending on surface modification of the clay layers, montmorillonite can be dispersed in a polymer matrix to form polymer-clay nanocomposite. Within the nanocomposite individual nm-thick clay layers become fully separated to form plate-like nanoparticles with very high (nm × μm) aspect ratio. Sigma-Aldrich, in collaboration with the Nanocor Corporation, offers a range of montmorillonite nanoclay products with different organic modifications optimized to be compatible with various polymer systems.

Halloysite Nanoclay

Halloysite is a naturally occurring aluminosilicate nanotube. The two-layer halloysite tubes are chemically similar to kaoline and have average dimensions of 15 × 1000 nm comparable to carbon nanotubes. Halloysite tubes are hollow and can be used for controlled delivery and release of drugs as well as nanocomposite and rheology modification applications.

Biomimetics

Biomimicry or biomimetics is the imitation of the models, systems, and elements of nature for the purpose of solving complex human problems. The terms *biomimicry* and *biomimetics* come from the Greek words bios, meaning life, and mimesis, meaning to imitate. A closely related field is bionics.

Over the last 3.6 billion years, nature has gone through a process of trial and error to refine the living organisms, processes, and materials on Earth. The emerging field of biomimetics has given rise to new technologies created from biologically inspired engineering at both the macro scale and nanoscale levels. Biomimetics is not a new idea. Humans have been looking at nature for answers to both complex and simple problems throughout our existence. Nature has solved many of today's engineering problems such as self-healing abilities, environmental exposure tolerance and resistance, hydrophobicity, self-assembly, and harnessing solar energy through the evolutionary mechanics of selective advantages.

One of the early examples of biomimicry was the study of birds to enable human flight. Although never successful in creating a "flying machine", Leonardo da Vinci (1452–1519) was a keen observer of the anatomy and flight of birds, and made numerous notes and sketches on his observations as well as sketches of "flying machines". The Wright Brothers, who succeeded in flying the first heavier-than-air aircraft in 1903, derived inspiration from observations of pigeons in flight.

Otto Schmitt, an American academic and inventor, coined the term *biomimetics* to describe the transfer of ideas from biology to technology. He developed the Schmitt trigger while attempting to replicate the biological system of nerve propagation. The term biomimetics entered Webster's Dictionary in 1974 and is defined as "the study of the formation, structure, or function of biologically produced substances and materials (as enzymes or silk) and biological mechanisms and processes (as protein synthesis or photosynthesis) especially for the purpose of synthesizing similar products by artificial mechanisms which mimic natural ones".

In 1960, the term *bionics* was coined by psychiatrist and engineer Jack Steele to mean "the science of systems which have some function copied from nature". *Bionics* entered the Webster dictionary in 1960 as "a science concerned with the application of data about the functioning of biological systems to the solution of engineering problems".

Bionic took on a different connotation when Martin Caidin referenced Jack Steele and his work in the novel *Cyborg* which later resulted in the 1974 television series *The Six Million Dollar Man* and its spin-offs. The term bionic then became associated with "the use of electronically operated artificial body parts" and "having ordinary human powers increased by or as if by the aid of such devices". Because the term *bionic* took on the implication of supernatural strength, the scientific community in English speaking countries largely abandoned it.

The term *biomimicry* appeared as early as 1982. Biomimicry was popularized by scientist and author Janine Benyus in her 1997 book *Biomimicry: Innovation Inspired by Nature*. Biomimicry is defined in the book as a "new science that studies nature's models and then imitates or takes inspiration from these designs and processes to solve human problems". Benyus suggests looking to Nature as a "Model, Measure, and Mentor" and emphasizes sustainability as an objective of biomimicry.

Nanobiomimicry

Biological imitation of nano and micro scale structures and processes is called nanobiomimicry. Nature provides a great variety of nano-sized materials that offer as potential templates for the creation of new materials, such as bacteria, viruses, diatoms, and biomolecules. Through the study of nanobiomimicry, key components of nanodevices like nanowires, quantum dots, and nanotubes have been produced in an efficient and simple manner when compared to more conventional lithographic techniques. Many of these biologically derived structures are then developed into applications for photovoltaics, sensors, filtration, insulation, and medical uses. The field of nanobiomimetics is highly multidisciplinary, and requires collaboration between biologists, engineers, physicists, material scientists, nanotechnologists and other related fields. In the past century, the growing field of nanotechnology has produced several novel materials and enabled scientists to produce nanoscale biological replicas.

Fabrication

Biomorphic mineralization is a technique that produces materials with morphologies and structures resembling those of natural living organisms by using bio-structures as templates for mineralization. Compared to other methods of material production, biomorphic mineralization is facile, environmentally benign and economic. Biomorphic mineralization makes efficient use of natural and abundant materials such as calcium, iron, carbon, phosphorus, and silicon with the capability of turning biomass wastes into useful materials. Templates derived from biological nanoparticles such as DNA, viruses, bacteria, and peptides can transform unordered inorganic nanoparticles into complex inorganic nanostructures. Biologically derived nanostructures are typically fabricated using either chemical or physical techniques. Typical chemical fabrication techniques are plasma spraying, plasma immersion ion implantation and deposition, sol–gel, chemical vapor deposition, physical vapor deposition, cold spraying and self-assembly. Physical modification techniques include laser etching, shot blasting, physical plating, and physical evaporation and deposition. Methods of fabrication with high throughput, minimal environmental damage, and low costs are highly sought after.

Biologically Inspired Engineering

The use of biomineralized structures is vast and derived from the abundance of nature. From studying the nano-scale morphology of living organisms many applications have been developed through multidisciplinary collaboration between biologists, chemists, bioengineers, nanotechnologists, and materials scientists.

Biomedicine

Mimicking the diving behaviour of animals, researchers have recently discovered that humans have a similar capacity to lower brain temperature and suppress metabolism for neuroprotection. This has now opened a real possibility of devising means for humans to sustain this state, not unlike the elusive and enigmatic feat of animal hibernation, e.g., lemurs (primates) and bears. This would have profound biomedical implications for healthcare and for treating an unmatched range and diversity of serious life-threatening clinical conditions, and in a fully personalized way, things like stroke, blood-loss, burns, cancer, chronic obesity, epileptic seizures, etc. An experimental trial, recently conducted in Sweden seemingly resulted in a sustainable variant of this state in a human breath-hold diver.

Nanowires, Nanotubes, and Quantum Dots

A virus is a nonliving particle ranging from the size of 20 to 300 nm capsules containing genetic material used to infect its host. The outer layer of viruses are remarkably robust and capable of withstanding temperatures as high as 60 °C and stay stable in a wide range of pH range of 2-10. Viral capsids can be used to create several nano device components such as nanowires, nanotubes, and quantum dots. Tubular virus particles such as the tobacco mosaic virus (TMV) can be used as templates to create nanofibres and nanotubes since both the inner and outer layers of the virus are charged surfaces and can induce nucleation of crystal growth. This was demonstrated though the production of platinum and gold nanotubes using TMV as a template. Mineralized virus particles have been shown to withstand various pH values by mineralizing the viruses with different materials such as silicon, PbS, and CdS and could therefore serve as a useful carriers of material. A spherical plant virus called cowpea chlorotic mottle virus (CCMV) has interesting expanding properties when exposed to environments of pH higher than 6.5. Above this pH, 60 independent pores with diameters about 2 nm begin to exchange substance with the environment. The structural transition of the viral capsid can be utilised in Biomorphic mineralization for selective uptake and deposition of minerals by controlling the solution pH. Applications include using the viral cage to produce uniformly shaped and sized quantum dot semiconductor nanoparticles through a series of pH washes. This is an alternative to the apoferritin cage technique currently used to synthesize uniform CdSe nanoparticles.

Such materials could also be used for targeted drug delivery since particles release contents upon exposure to specific pH levels.

Display Technology

Morpho butterfly wings contain microstructures that create its colouring effect through structural colouration rather than pigmentation. Incident light waves are reflected at specific wavelengths to create vibrant colours due to multilayer interference, diffraction, thin film interference, and scattering properties. The scales of these butterflies consist of microstructures such as ridges, cross-ribs, ridge-lamellae, and microribs that have been shown to be responsible for colouration. The structural colour has been simply explained as the interference due to alternating layers of cuticle and air using a model of multilayer interference. The same principles behind the colouration of soap bubbles apply to butterfly wings. The colour of butterfly wings is due to multiple instances of constructive interference from structures such as this. The photonic microstructure of butterfly wings can be replicated through biomorphic mineralization to yield similar properties. The photonic microstructures can be replicated using metal oxides or metal alkoxides such as titanium sulfate ($TiSO_4$), zirconium oxide (ZrO_2), and aluminium oxide (Al_2O_3). An alternative method of vapor-phase oxidation of SiH4 on the template surface was found to preserve delicate structural features of the microstructure. Now, companies like Qualcomm are specialising in creating colour displays with low power consumption based on these principles.

Additional Examples

Researchers studied the termite's ability to maintain virtually constant temperature and humidity in their termite mounds in Africa despite outside temperatures that vary from 1.5 °C to 40 °C (35 °F to 104 °F). Researchers initially scanned a termite mound and created 3-D images of the mound structure, which revealed construction that can influence human building design. The Eastgate Centre, a mid-rise office complex in Harare, Zimbabwe, stays cool without air conditioning and uses only 10% of the energy of a conventional building its size.

Modelling echolocation in bats in darkness has led to a cane for the visually impaired. Research at the University of Leeds, in the United Kingdom, led to the UltraCane, a product formerly manufactured, marketed and sold by Sound Foresight Ltd.

Janine Benyus refers in her books to spiders that create web silk as strong as the Kevlar used in bulletproof vests. Engineers could use such a material—if it had a long enough rate of decay—for parachute lines, suspension bridge cables, artificial ligaments for medicine, and other purposes.

Other research has proposed adhesive glue from mussels, solar cells made like leaves, fabric that emulates shark skin, harvesting water from fog like a beetle, and more. Nature's 100 Best is a compilation of the top hundred different innovations of animals, plants, and other organisms that have been researched and studied by the Biomimicry Institute.

A display technology based on the reflective properties of certain morpho butterflies was commercialized by Qualcomm in 2007. The technology uses Interferometric Modulation to reflect light so only the desired colour is visible in each individual pixel of the display.

Biomimicry may also provide design methodologies and techniques to optimize engineering products and systems. An example is the re-derivation of Murray's law, which in conventional form determined the optimum diameter of blood vessels, to provide simple equations for the pipe or tube diameter which gives a minimum mass engineering system.

In structural engineering, the Swiss Federal Institute of Technology (EPFL) has incorporated biomimetic characteristics in an adaptive deployable "tensegrity" bridge. The bridge can carry out self-diagnosis and self-repair.

The extremely tough and mechanically versatile insect cuticle, inspired the scientists from the Wyss Institute for Biologically Inspired Engineering at Harvard University, who developed the Shrilk family of biomimetic plastics, based in the components and design of the insect skin.

The Bombardier beetle's powerful repellent spray inspired a Swedish company to develop a "micro mist" spray technology, which is claimed to have a low carbon impact (compared to aerosol sprays). The beetle mixes chemicals and releases its spray via a steerable nozzle at the end of its abdomen, stinging and confusing the victim.

Holistic planned grazing, using fencing and/or herders, seeks to restore grasslands by carefully planning movements of large herds of livestock to mimic the vast herds found in nature where grazing animals are kept concentrated by pack predators and must move on after eating, trampling, and manuring an area, returning only after it has fully recovered. Developed by Allan Savory, this method of biomimetic grazing holds tremendous potential in building soil, increasing biodiversity, reversing desertification, and mitigating global warming, similar to what occurred during the past 40 million years as the expansion of grass-grazer ecosystems built deep grassland soils, sequestering carbon and cooling the planet.

Biomimetic Design of Molecules

Artificial photosynthesis for fuel production is a quite recent field. Our knowledge about photosynthesis and the natural H2 forming reactions increases rapidly and during the last 10 years the three-dimensional structures of the photosynthetic reaction centres Photosystem I and II and many Hydrogenases have been solved. This sophisticated knowledge is provided by the photosynthesis group and is key for the biomimetic approach applied by the synthetic chemists in the Consortium for Artificial Photosynthesis. There have been many attempts to model crucial physical and chemical elementary steps in these enzymes by bio-mimetic chemistry. However, there have been few attempts to use the biomimetic approach to actually produce a fuel from solar energy similar to what is done in photosynthetic organisms.

Biomimetic Nanomaterials

Reference to biological models inspiring engineers to develop new materials and technologies is based on the assumption that nature, in many billion years of evolution, has created optimal living structures that are far more effective and endurable than any man-made structures. For example, the study of the "lotus effect" , i.e. the property of lotus leaves to resist wetting and dirt due to their micro- and nanostructured surface has led to the development of waterproof paints and textiles. Polymer nanofibres whose strength is comparable to that of steel were inspired by an example from biology, namely the cobweb whose threads can withstand three times greater strain than steel wire with equal diameter. Burdock fruit became a prototype for the synthetic adhesive material Velcro, the basis of widely used hook-and-loop fasteners.

Many biomolecules are capable of self-assembly in regular structures. For example, under polymerisation conditions the contractile protein actin forms filaments 7 nm thick, and tubulin forms microtubes 25 nm in diameter. Using this self-assembling mechanism and the biostructures as matrices , nanoconductors and nanotubes can be created through the deposition of metal monolayers onto biopolymers. Complementarity that underlies the double-stranded DNA assembly is used in the development of new DNA-based nanomaterials.

Knowing the structure and functions of biological molecules, we can synthesise hybrid molecules, including peptides, lipids and organic polymers, and develop biomimetic nanofibres, bioinorganic composites and nanoporous coatings for tissue engineering.

Biomimetic nanoparticles are under active development. For example, ferritin, a protein that transports and stores iron in an organism, forms nanocavities with an inner diameter of 8 nm. Magnetic nanoparticles of iron oxide and cobalt oxide with a mean size of 6 nm can be encapsulated in these nanocavities. Other approaches employ "growing" nanoparticles of specific sizes inside bacteria or the biomass of plants (oats, wheat or alfalfa). Metal salts are added to these biological objects, and after biocatalitic reduction to metals they form nanoparticles. Methods of growing metallic nanoparticles in plants by adding metal salts to the irrigation water have been described. Nanoparticles are formed in the stems and other parts of plants and can be extracted. Proteins involved in reducing reactions define the size of the formed nanoparticles. In some cases peptide sequences responsible for catalysis have been identified, which allowed them to be used as ring peptides for the in vitro development of nanoparticles. Nanoparticles can also be created using viral shells (capsids). Viral capsid proteins are assembled to form regular-shaped three-dimensional structures with a hollow space inside where the virus genome is packed. High regularity calibrated metallic nanoparticles and nanocomposites can be assembled both inside a capsid and on its surface. Biomimetic synthesis of nanoparticles has certain advantages; for example, it is carried out under milder conditions compared to the physical and chemical production of nanoparticles. In the large-scale production of nanoparticles this feature makes it possible to minimise adverse environmental effects.

Use of Nanotechnology in Developing Countries

Where wealth is concerned, the least developed countries present the hardest case. Can a capability as advanced as nanotechnology, based on molecular machinery, be of use in the Third World? The answer must be yes. Agriculture is the backbone of Third World economies today, and agriculture is based on the naturally occurring molecular machines in wheat, rice, yams, and the like.

The Third World is short on equipment and skills. (It often has governmental problems as well, but that is another story.) Molecular manufacturing can make equipment inexpensive enough for the poor to buy or for aid agencies to give away. This includes equipment for making more equipment, so dependency could be reduced. As for skills, basic molecular manufacturing will require little labour of any kind, and a little skill will go a long way. As the technology advances, more and more of the products can be easy-to-use smart materials.

Molecular manufacturing will enable the poorest countries to bypass the difficult and dirty process of the industrial revolution. It can make products that are less expensive and easier to use than yams or rice or goats or water buffalo. And with products like cheap supercomputers with huge databases of writing and animation viewed through 3-D colour displays, it can even help spread knowledge.

Nanotechnology's role in helping the poorest nations won't be on the minds of the first developers-they'll be in government and commercial labs in the wealthiest nations, pursuing problems of concern to people there. History, though, is full of unintended consequences, and some are for the better.

Construction and Housing

Building large objects is basic to solving problems of housing and transportation. Smart materials can help. Today, buildings are expensive to construct, expensive to replace, and expensive to make fireproof, tornado-proof, earthquake-proof, and so forth. Making buildings tall is expensive; making walls soundproof is expensive; building underground is expensive. Efforts to relieve city congestion often founder on the high cost of building subways, which can amount to hundreds of millions of dollars per mile. Building codes and politics permitting, nanotechnology will make possible revolutions in the construction of buildings. Superior materials will make it easy to construct tall (or deep) buildings to free up land, and strong buildings that can ride out the greatest earthquake without harm. Buildings can be made so energy-efficient and so good at using the solar energy falling on them that most are net energy producers. What is more, smart materials can make it easy to build and modify complex structures, such as walls full of windows, wiring, plumbing, data networks, and the like. For a concrete example that shows the principle, let's picture what smart pipes could be like.

Let's say that you want to install a fold-down sink in the corner of your bedroom. The new materials make fold-down sinks practical, and in a house made of advanced smart materials, just sticking one on the wall would be enough-the plumbing would rearrange itself. But this is an old, pre-breakthrough house, so the sink is a retrofit. To do this home-handiwork project, you buy several boxes full of inexpensive tubing, T-joints, valves, and fixtures in a variety of sizes, all as light as wood veneer and feeling like soft rubber.

The biggest practical problem will be to make a hole from an existing water pipe and drainpipe to where you want the sink. Molecular manufacturing can provide excellent power tools to make the holes, and smart paint and plaster to cover them again, but the details depend on how your house is built.

The smart plumbing system does help, of course. If you want to run the drain line through the attic, built-in pumps will make sure that the water flows properly. The flexibility of the pipes makes it much easier to run them around curves and corners. Low-cost power makes it practical for the sink to have a flow-through water heater, so you only need to run a cold-water pipe to have both hot and cold water. All the parts go together as easily as a child's blocks, and seem about as flimsy and likely to leak. When you turn it on, though, the microscopic components of the pipes lock together and become as strong as steel. And smart plumbing doesn't leak.

If your house were made of smart materials, like most of the housing in the Third World these days, life would have been easier. Using a special trowel, wall structures would be reworked like soft clay, doing their structural job all the while. Setting up a plumbing system from scratch with this stuff is easy, and hard to do wrong. Drinking water pipes won't connect to wastewater pipes, so drinking water can't be accidentally contaminated. Drains won't clog, because they can clean themselves better than a rotary steel blade ever could. If you run enough pipes from everything to everything else, built-in pumps will make sure that water flows in the right direction with adequate pressure. Smart plumbing is one example of a general pattern. Molecular manufacturing can eventually make complex products at low cost, and those complex products can be simpler to use than anything we have today, freeing our attention for other concerns. Buildings can become easy to make and easy to change. The basic conveniences of the modern world, and more, can be carried to the ends of the earth and installed by the people there to suit their tastes.

Food

Worldwide food production has been outpacing population growth, yet hunger continues. In recent years, famine has often had political roots, as in Ethiopia where the rulers aim to starve opponents into submission. Such problems are beyond a simple technological solution. To avoid getting headaches, we'll also ignore the politics of farm price-support programs, which raise food prices while people are going hungry. All we can suggest here is a way to provide fresh food at lower cost with reduced environmental impact.

For decades, futurists have predicted the coming of synthetic foods. Some sort of molecular-manufacturing process could doubtless make such things with the usual low costs, but this doesn't sound appetizing, so we'll ignore the idea.

Most agriculture today is inefficient-an environmental disaster. Modern agriculture is famed for wasting water and polluting it with synthetic fertilizers, and for spreading herbicides and pesticides over the landscape. Yet the greatest environmental impact of agriculture is its sheer consumption of land. In the American East, ancient forests disappeared under the ax, in part to supply wood, in part to clear land. The prairies of the West disappeared under the plow.

Around the world, this trend continues. The technology of the ax, the fire, and the plow is chiefly responsible for the destruction of rain forests today. A growing population will tend to turn every productive ecosystem into some sort of farmland or grazing land, if we let it. No technological fix can solve the long-term problem of population growth. Nonetheless, we can roll back the problem of the loss of land, yet increase food supplies. One approach is intensive greenhouse agriculture.

Every kind of plant has its optimum growing conditions, and those conditions are far different from those found in most farmland during most of the year. Plants growing outdoors face insect pests, unless doused with pesticide, and low levels of nutrients, unless doused with fertilizer. In greenhouses patrolled by "nanoflyswatters" able to eliminate invading insects, plants would be protected from pests and could be provided with nutrients without contaminating groundwater or runoff. Most plants prefer higher humidity than most climates provide. Most plants prefer higher, more uniform temperatures than are typically found outdoors. What is more, plants thrive in high levels of carbon dioxide. Only greenhouses can provide pest protection, ample nutrients, humidity, warmth, and carbon dioxide all together and without re-engineering the Earth.

Taken together, these factors make a huge difference in agricultural productivity. Experiments with intensive greenhouse agriculture, performed by the Environmental Research Lab in Arizona, show that an area of 250 square metres-about the size of a tennis court-can raise enough food for a person, year in and year out.

With molecular manufacturing to make inexpensive, reliable equipment, the intensive labour of intensive agriculture can be automated. With technology like the deployable "tents" and smart materials we have described, greenhouse construction can be inexpensive. Following the standard argument, with equipment costs, labour costs, materials costs, and so forth, all expected to be low, greenhouse-grown foods can be inexpensive.

What does this mean for the environment? It means that the human race could feed itself with ordinary, naturally grown, pesticide-free foods while returning more than 90 percent of today's agricultural land to wilds. With a generous five hundred square metres per person, the U.S. population would require only 3 percent of present U.S. farm acreage, freeing 97 percent for other uses, or for a gradual return to wilderness.

When farmers are able to grow high-quality foodstuffs inexpensively, in a fraction of the room that they require today, they will find more demand for their land to be tended as a park or wilderness than as a cornfield. Farm journals can be expected to carry articles advising on techniques for rapid and esthetic restoration of forest and grassland, and on how best to accommodate the desires of the discriminating nature lover and conservationist. Even "unpopular" land will tend to become popular with people seeking solitude.

The economics of assembler-based manufacturing will remove the incentive to make greenhouses cheap, ugly, and boxy; the only reason to build that way today is the high cost of building anything at all. And while today's greenhouses suffer from viral and fungal infestations, these could be eradicated from plants in the same way they would be from the human body, as will be described later. A problem faced by today's greenhouses-overheating-could be dealt with by using heat exchangers, thereby conserving the carefully balanced inside atmosphere. Finally, if it should turn out that a little bit of bad weather improves the taste of tomatoes, that, too, could be provided, since there would be no reason to be fanatical about sheer efficiency.

Communications

Today, telecommunications systems have sharply limited capacity and are expensive to expand. Molecular manufacturing will drop the price of the "boxes" in telecommunications systems-things such as switching systems, computers, telephones, and even the fabled videophone. Cables made of smart materials can make these devices easy to install and easy to connect together.

Regulatory agencies willing, you might someday be able to buy inexpensive spools of material resembling kite string, and other spools of material resembling tape, then use them to join a world data network. Either kind of strand can configure its core into a good-quality optical fibre, with special provisions for going around bends. When rubbed together, pieces of string will fuse together, or fuse to a piece of tape. Pieces of tape do likewise. To hook up to the network, you run string or tape from your telephone or other data terminal to the nearest point that is already connected. If you live deep in a tropical rain forest, run a string to the village satellite link.

These data-cable materials include amplifiers, nanocomputers, switching nodes, and the rest, and they come loaded with software that "knows" how to act to transmit data reliably. If you're worried that a line may break, run three in different directions. Even one line could carry far more data than all the channels in a television cable put together.

Transportation

Getting around quickly requires vehicles and somewhere for them to travel. The old 1950s vision of private helicopters would be technically possible with inexpensive, high-quality manufacturing, cheap energy, and a bit of improvement in autopilots and air-traffic control-but will people really tolerate that much junk roaring across the sky? Fortunately, there is an alternative both to this and to building ever more roads.

Going Underground

Near the surface of the Earth, there is as much room underground as there is above it. This is usually ignored, because the room is full of dirt, rock, pressurized water, and the like. Digging is expensive. Digging long, deep tunnels is even more expensive. This expense, however, is mostly in the cost of equipment, materials, and energy. Tunneling machines are in wide use today, and molecular manufacturing can make them more efficient and less expensive. The energy to operate them will be no great problem, and smart materials can line tunnels as fast as they are dug, with little or no labour. Nanotechnology will open the low frontier.

With a little care, the environmental impact of a deep tunnel can be trivial. Instead of solid rock far below the surface, there is rock with a sealed tunnel running through it. Nothing nearby need be disturbed. Tunnels avoid both the aesthetic impact of a sky full of noisy aircraft and the environmental impact of paving strips of landscape. This will make them less expensive than roads, and they can, if desired, be more common than roads in the developed world today. They will even permit faster transportation.

Taking the Subway

Japan and Germany are actively developing magnetic trains, like those in the Desert Rose scenario. These avoid the limitations of steel wheels on steel rails by using magnetic forces to "fly" the train along a special track. Magnetic trains can reach aircraft speeds at ground level. On long runs through evacuated tunnels, they can reach spacecraft speeds, travelling global distances in an hour or so (less, if passengers are willing to tolerate substantial acceleration).

Systems like this can give "taking the subway" a new meaning. Local transportation would be at fast automotive speeds, but long-distance transportation would be faster than the Concorde. With superconducting electrical systems, fast subways would be more energy efficient than today's slow mass transit.

Getting Your Car

For decades, people have proposed replacing automobiles with some form of mass-transportation system, and it seems that cost revolutions (including inexpensive tunneling) may finally make this practical. Before junking the car, though, it's worth seeing how it might be improved. Molecular manufacturing can make almost anything better. Automobiles can be made stronger and safer, lighter, higher performance, and higher efficiency, while getting excellent mileage and burning clean, inexpensive fuels, perhaps in fuel cells powering quiet electric motors. Using aerodynamic forces to hold the car to the road, there's no reason why a comfortable passenger car shouldn't be able to deliver uncomfortable, drag-racer acceleration.

To imagine a cheap car built with molecular manufacturing, first imagine loading it with all the attractive features that you've ever heard proposed. This includes everything from today's self-adjusting seats and mirrors, excellent sound systems, and specially tuned steering and suspension systems, through automated navigation displays, emergency braking, and reliable super-duper airbags.

Now, instead of just having the position of the seats, mirrors, and so forth adjust to a driver, as some cars do today, our smart-material car can also adjust its size, shape, and colour, facing owners with choices such as, "What should our car look like for this occasion?" Those seeking an image of solid conservatism and wealth won't drive such cheap cars; they will risk their necks in a certified antique car, made from the traditional steel, paint, and rubber. If environmental regulations permit it, the car might even have a genuine gasoline-burning engine. The latter can no doubt be cleaned up by fancy nanotechnology-based emission-control systems.

Opening the Space Frontier

Our transportion system today effectively ends in the upper atmosphere. Travel beyond still takes the form of "historic missions." There is no reason for this situation to continue for long, once molecular manufacturing becomes well established.

The cost of spaceflight is high because spacecraft are huge, fragile things, made in such small numbers that they're almost hand-crafted.

Molecular manufacturing will replace today's delicate monsters with rugged, mass-produced vehicles (which, with greater efficiency, needn't be so large). The vehicles will cost little, but the energy? Today, the energy cost of a ticket to orbit in an efficient vehicle would be less than one hundred dollars. Low cost vehicles and energy will drop the total cost to a fraction of this.

We will know that spaceflight has become inexpensive when people see the Earth as just a small part of the world, and understand in their bones that space resources make continued exploitation of Earth's resources unnecessary. In the long run, efficient, clean, low-cost manufacturing can transform the way human beings affect the Earth by their presence. Even stay-at-home humans will be better able to heal the damage they have done.

Ecosystem Protectors

In many parts of the world, native species have been driven to extinction by rats, pigs, and other imported species, and others are endangered and fighting for their lives. Biological controls-fighting fire with fire-have advantages: organisms are small, selective, and inexpensive. These advantages will eventually be shared by devices made using molecular manufacturing, which avoid the disadvantages of importing and releasing yet more uncontrollable, breeding, spreading species. Alan Liss spoke of using nanotechnological devices to help restore ecosystems at a chemical level. A similar idea can be applied at a biological level.

The challenge-and it is huge-would be to develop insect-size or even microbe-size devices that could serve as selective, mobile, mechanical flyswatters or weed pullers. These could do what biological controls do, but would be unable to replicate and spread. Let's call devices of this sort "ecosystem protectors." They could keep aggressive imported species out, saving native species from extinction. To a human being or an ordinary organism, an ecosystem protector would seem like just one more of the many billions of different kinds of bugs and microbes in the ecosystem-small things going about their own business, with no tendency to bite.

They might be detectable, but only if you sorted through a lot of dirt and looked at it through a microscope, because they wouldn't be very common. They would have just one purpose: to notice when they bumped into a member of an imported species on the "not welcome here" list, and then either to eliminate it or to ensure, at least, that it couldn't reproduce.

Natural organisms are often very finicky about which species they attack. These ecosystem protectors could be equally finicky about which species they approach, and then, before attacking, could do a DNA analysis to be sure. It would be simplest (especially in the beginning while we're still learning) to limit each kind of defender to monitoring only one imported species. Each unit of a particular kind of ecosystem-defender device would be identical, built with precision by a special-purpose molecular-manufacturing setup. Each would last for a certain time, then break down. Each kind can be tested in a terrarium, then a greenhouse, then a trial outdoors ecosystem, keeping an eye on their effects at each stage until one gains the confidence for larger scale use.

"Larger scale" could still be quite limited, if they aren't designed to travel very far. This built-in obsolescence limits both how long each device can operate and how far it can move: getting control of the structure of matter includes making nanomachines work where they're wanted and not work elsewhere.

The agricultural industry today manufactures and distributes many thousands of tons of poisonous chemicals to be sprayed on the land, typically in an attempt to eliminate one or a few species of insect.

Ecosystem protectors could also be used to protect these agricultural monocultures, field by field, with far less harm to the environment than today's methods. They could likewise be used in the special ecosystems of intensive greenhouse agriculture.

Unlike chemicals sprayed into the environment, these ecosystem protectors would be precisely limited in time, space, and effect. They neither contaminate the groundwater nor poison bees and ladybugs. In order to weed out imported organisms and bring an ecosystem back to its natural balance, ecosystem protectors would not have to be very common-only common enough for a typical imported organism to encounter one once in a lifetime, before reproducing.

Even so, as the ecosystem protectors wear out and stop working, they would present a small-scale problem of solid-waste disposal.

With the exercise of some clever design, all the machinery of ecosystem protectors might be made of reasonably durable yet biodegradable materials or (at worst) materials no more harmful than bits of grit and humus in the soil. So their remains would be like the shells of diatoms, or bits of lignin from wood, or like peculiar particles of clay or sand.

Alternatively, we might develop other mobile nanomachines to find and collect or break down their remains. This strategy starts to look like setting up a parallel ecosystem of mobile machines, a process that could be extended to supplement the natural cleansing processes of nature in many ways. Each step in this direction will require caution, but not paranoia: there need be no toxic chemicals here, no new creatures to spread and run wild. Missteps will have the great virtue of being reversible.

If we decide that we don't like the effects of some particular variety of ecosystem protector or cleanup machine, we could simply stop manufacturing that kind. We could even retrieve those that had already been made and dispersed in the environment, since their exact number is known, along with which patch of ground each is patrolling.

If the making and monitoring of ecosystem protectors seems a lot of trouble to go to just to weed out nonnative species, consider this example of the environmental destruction such species can cause. Sometime before World War II, a South African species of fire ant was accidentally imported into the United States.

Today, infested areas can have up to five hundred of these ants per square foot. The National Audubon Society-a strong opponent of irresponsible use of pesticides-had to resort to spraying its refuge islands near Corpus Christi when they found these ants destroying over half the hatchlings of the brown pelican, an endangered species.

In Texas, it's been shown that the new ants are killing off native ant species-reducing biodiversity. The USDA's Sanford Porter states that due to them, "Texas may be in the midst of a genuine biological revolution." The ants are heading west, and have established a beachhead in California. Without ecosystem protectors or something much like them, ecologies around the world will continue to be threatened by unnatural invasions. Our species opened the new invasion routes, and it's our responsibility to protect native species made newly vulnerable by them.

Mending the Land

Today, most people are far from the land, tied up in turning the wheels of 20th century industry. In the years to come, those wheels will be replaced by molecular systems that do most of their turning by themselves. The pressure to destroy the land will be less. Time available to help heal the land will be greater. Surely more energy will flow in this direction. To mend ruined landscapes will require skill and effort. Ecosystem defenders can do flyswatting and weedpulling jobs no humans ever could, but there will also be jobs of shaping, planting, and nurturing. The land has been torn by machines guided by hasty hands, almost overnight. It can gradually be restored by patient hands, whether bare, gloved, or guiding machines able to reshape a ravaged mountain without turning the soil.

The green wealth that can be brought by nanotechnology has raised high hopes among some environmentalists. Again writing in Whole Earth Review, Terence McKenna suggests it "would tend to promote... a sense of the unity and balance of nature and of our own human position within that dynamic and evolving balance." Perhaps people will learn to value nature more deeply when they can see it more clearly, with eyes unclouded by grief and guilt.

Ending Chemical Pollution, Cutting Resource Consumption

We've already seen how molecular manufacturing can provide clean solar energy without paving over desert ecosystems, and how clean energy and common materials can be turned into abundant, efficient goods, also cleanly. With care, sources of chemical pollution-even of excess carbon dioxide-can, step by step, be eliminated. This includes the pollutants responsible for acid rain, as well as ozone-destroying gases, greenhouse gases, oil spills, and toxic wastes.

In each case, the story is about the same. Acid rain mostly results from burning dirty fuels containing sulfur, and from burning cleaner fuels in a dirty way, producing nitrogen oxides. We've seen how molecular manu-facturing can make solar cells cheap enough and rugged enough to use as road surfaces.

With green wealth, we can make clean fuels from solar energy, air, and water; consuming these fuels in clean nanomechanical systems would just return to the air exactly the materials taken from it, along with a little water vapor. Fuels are made, fuels are consumed, and the cycle produces no net pollution. With cheap solar fuels, coal and petroleum can be replaced, ignored, left in the ground. When petroleum is obsolete, oil spills will vanish.

The greenhouse gas of greatest concern is carbon dioxide, and its main source is the burning of fossil fuels. The above steps would end this. The release of other gases, such as the chlorofluorocarbons(CFCs) used in foaming plastics, is often a side effect of primitive manufacturing processes: foaming plastic will hardly be a popular activity in an era of molecular manufacturing. These materials can be replaced or controlled-and they include the gases most responsible for ozone depletion.

The chief threats to the ozone layer are those same CFCs, used as refrigerants and solvents. Molecular manufacturing will use solvents sparingly (mostly water), and can recycle them without dumping any. CFC refrigerants can be replaced even with current technology, at a cost; with nanotechnology, that cost will be negligible.

Toxic wastes generally consist of harmless atoms arranged into noxious molecules; the same is true of sewage. With inexpensive energy and equipment able to work at the molecular level, these wastes can be converted into harmless forms. Many need never be produced in the first place. Other toxic wastes contain toxic elements, such as lead, mercury, arsenic, and cadmium. These elements come from the ground, and are best returned to the location and condition in which they were found. With nanotechnology, moreover, there will be little reason to dig them up in the first place. Nanotechnology will be able to break materials down to simple molecules and build them back up again. Need it be said that this will permit complete recycling?

It is fair to say that eliminating these sources of pollution would be a major improvement. There doesn't seem to be much more to say, aside from the usual caveats: "Not immediately," "Not all at once," and "Not on a predictable schedule." No one wants to make and dump wastes; they want something else, and get wastes as by-products. With a better way to get what people want, dumping wastes can be stopped. People will also be able to get what they want while reducing their resource consumption. As materials grow stronger, they can be used more sparingly. As machines grow more perfect-in their motors, bearings, insulation, computers-they will grow more efficient. Materials will be needed to make things, and energy will be needed to run them, but in smaller amounts. What is more, nanotechnology will be the ultimate recycling technology.

Objects can be made extremely durable, decreasing the need for recycling; alternatively, objects can be made genuinely biodegradable, designed at the molecular level to decompose after use, leaving humus and mineral grit; alternatively, they can be made of microscopic snap-together pieces, making objects as recyclable as structures built and rebuilt out of a child's blocks; finally, even objects not designed for recycling can be taken apart into simple molecules and recycled regardless. Each approach has different advantages and costs, and each makes current garbage problems go away.

Cleaning up the Twentieth Century Mess

Still, even after twentieth-century industry is history, its toxic residues will remain. Cleaning up waste dumps with today's technology has proved so expensive and ineffective that many in the field have all but given up hope of really solving the problem. What can be done with post-breakthrough technologies?

Cleansing Soil and Water

Nanotechnology can help with the cleanup of these pollutants. Living organisms clean the environment, when they can, by using molecular machinery to break down toxic materials. Systems built with nanotechnology will be able to do likewise, and to deal with compounds that aren't biodegradable.

Figure : *Environmental Cleanup: By changing the way materials and products are made, molecular manufacturing will free up land formerly used for industrial plants. Toxic materials could be removed from contaminated soil using solar power as the energy source, and the cleanup device and any collected residues could later be carted away.*

Alan Liss is director of research for Ecological Engineering Associates, a company that uses knowledge of how natural ecosystems function to address environmental problems such as wastewater treatment. He explains how cleanup could work: "The more we learn about the ecosystem, the more we find that functions are managed by particular organisms or groups of organisms.

Nanotech 'managers' might be able to step in when the natural managers are not available, thereby having a particular ecological activity occur that otherwise wouldn't have happened. A nanotech manager might be used for remediation in a situation where toxicants have destroyed some key members of a particular ecosystem-some managerial microbes, for example. Once the needed activities are reinitiated, the living survivors of the stressed ecosystem can jump in and continue the ecosystem recovery effort."

To see how nanomachines could be used to clean up pollution, imagine a device made of smart materials and roughly resembling a tree, once it has been delivered and unfolded. Above ground are solar-collecting panels; below ground, a branching system of rootlike tubes reaches a certain distance into the soil. By extending into a toxic waste dump, these rootlike structures could soak up toxic chemicals, using energy from the solar collectors to convert them into harmless compounds. Rootlike structures extending down into the water table could do the same cleanup job in polluted aquifers.

Cleansing the Atmosphere

Most atmospheric pollutants are quickly washed out by rain (turning them into soil-and water-pollution problems), but some air pollutants are longer lasting. Among these are the chlorine compounds attacking the ozone layer that protects the Earth from excessive ultraviolet radiation. Since 1975, observers have recorded growing holes in the ozone layer: at the South Pole, the hole can reach as far as the tips of South America, Africa, and Australia. Loss of this protection subjects people to an increased risk of skin cancer and has unknown effects on ecosystems. The new technology base will be able to stop the increase in ozone-destroying compounds, but the effects would linger for years. How might this problem be reversed more rapidly?

Thus far, we've talked about nanotechnology in the laboratory, in manufacturing plants, and in products for direct human use. Molecular manufacturing can also make products that will perform some useful temporary function when tossed out into the environment. Getting rid of ozone-destroying pollutants high in the stratosphere is one example. There may be simpler approaches, without the sophistication of nanotechnology, but here is one that would work to cleanse the stratosphere of chlorine: Make huge numbers of balloons, each the size of a grain of pollen and light enough to float up into the ozone layer. In each, place a small solar-power plant, a molecular-processing plant, and a microscopic grain of sodium. The processing plant collects chlorine-containing compounds and separates out the chlorine. Combining this with the sodium makes sodium chloride-ordinary salt. When the sodium is gone, the balloon collapses and falls. Eventually, a grain of salt and a biodegradable speck fall to Earth, usually at sea. The stratosphere is soon clean.

A larger problem (with a ground-based solution) is climatic change caused by rising carbon dioxide (CO2) levels. Global warming, expected by most climatologists and probably under way today, is caused by changes in the composition of Earth's atmosphere. The sun shines on the Earth, warming it. The Earth radiates heat back into space, cooling. The rate at which it cools depends on how transparent the atmosphere is to the radiation of heat.

The tendency of the atmosphere to hold heat, to block thermal radiation from escaping into space, causes what is called the "greenhouse effect." Several gases contribute to this, but CO2 presents the most massive problem. Fossil fuels and deforestation both contribute. Before the new technology base arrives, something like 300 billion tons of excess CO2 will likely have been added to the atmosphere.

Small greenhouses can help reverse the global greenhouse effect. By permitting more efficient agriculture, molecular manufacturing can free land for reforestation, helping to repair the devastation wrought by hungry people. Growing forests absorb CO2.

If reforestation is not fast enough, inexpensive solar energy can be applied to remove CO2 directly, producing oxygen and glossy graphite pebbles. Painting the world's roads with solar cells would yield about four trillion watts of power, enough to remove CO2 at a rate of 10 billion tons per year. Temporarily planting one-tenth of U.S. farm acreage with a solar cell "crop" would provide enough energy to remove 300 billion tons in five years; winds would distribute the benefits worldwide. The twentieth century insult to Earth's atmosphere can be reversed by less than a decade of twenty-first century repair work. Ecosystems damaged in the meantime are another matter.

Orbital Waste

The space near Earth is being polluted with small orbiting projectiles, some as small as a pin. Most of the debris is floating fragments of discarded rocket stages, but it also includes gloves and cameras dropped by astronauts. This is not a problem for life on Earth, but it is a problem as life begins its historic spread beyond Earth-the first great expansion since the greening of the continents, long ago. Orbiting objects travel much faster than rifle bullets, and energy increases as the square of speed. Small fragments of debris in space can do tremendous damage to a spacecraft, and worse-their impact on an spacecraft can blast loose yet more debris. Each fragment is potentially deadly to a spacefaring human crossing its path. Today, the tiny fraction of space that is near Earth is increasingly cluttered. This litter needs to be picked up. With molecular manufacturing, it will be possible to build small spacecraft able to maneuver from orbit to orbit in space, picking up one piece of debris after another. Small spacecraft are needed, since it makes no sense to send a shuttle after a scrap of metal the size of a postage stamp. With these devices, we can clean the skies and keep them hospitable to life.

The Nuclear Waste

We've spoken of waste that just needs molecular changes to make it harmless, and toxic elements that came from the ground, but nuclear technology has created a third kind of waste. It has converted the slow, mild radioactivity of uranium into the fast, intense radioactivity of newly created nuclei, the products of fission and neutron bombardment. No molecular change can make them harmless, and these materials did not come from the ground. The products of molecular manufacturing could help with conventional approaches to dealing with nuclear waste, helping to store it in the most stable, reliable forms possible-but there is a more radical solution.

Even before the era of the nuclear reactor and the nuclear bomb, experimenters made artificially radioactive elements by accelerating particles and slamming them into nonradioactive targets. These particles travelled fast enough to penetrate the interior of an atom and reach the nucleus, joining it or breaking it apart.

The entire Earth is made of fallout from nuclear reactions in ancient stars. Its radioactivity is low because so much time has passed-many half-lives, for most radioactive nuclei. "Kicking" these stable nuclei changes them, often into a radioactive state. But kicking a radioactive nucleus has a certain chance of turning it into a stable one, destroying the radioactivity. By kicking, sorting, and kicking again, an atom-smashing machine could take in electrical power and radioactive waste, and output nothing but stable, nonradioactive elements, identical to those common in nature. Don't recommend this to your congressman-it would be far too expensive, today-but it will some day be practical to destroy the radioactivity of the twentieth-century's leftover nuclear waste.

Nanotechnology cannot do this directly, because molecular machines work with molecules, not nuclei. But indirectly, by making energy and equipment inexpensive, molecular manufacturing can give us the means for a clean, permanent solution to the problem of wastes left over from the nuclear era.

A Wealth of Garbage

Shortages often spur environmental damage. Faced with a food shortage, herdsmen can graze grasslands down to bare dirt. Faced with an energy shortage, industrial countries can approve destructive projects. The growth of population and the consumption of resources by twentieth-century industry have placed growing pressures on Earth's ability to support us in the manner to which we have become accustomed.

The resource problem will look quite different in the twenty-first century, with a new technology base. Today, we cut trees and mine iron for our structures. We pump oil and mine coal for our energy. Even cement is born in the flames of burning fossil fuels. Almost everything we build, almost every move we make, consumes something ripped from the Earth. This need not continue.

Our civilization uses materials for many things, but mainly to make things with a certain size, shape, and strength. These structural uses include everything from fibres in clothing to paving in roads, and most of the mass of furniture, walls, cars, spacecraft, computers-indeed, most of the mass of almost every product we build and use. The best structural materials use carbon, in forms like diamond and graphite.

With elements from air and water, carbon makes up the polymers of wool and polyester, and of wood and nylon. A twenty-first-century civilization could mine the atmosphere for carbon, extracting over 300 billion tons before lowering the CO2 concentration back to its natural, pre-industrial level. For a population of 10 billion, this would be enough to give every family a large house with lightweight but steel-strong walls, with 95 percent left over. Atmospheric garbage is an ample source of structural materials, with no need to cut trees or dig iron ore.

Plants show that carbon can be used to build solar collectors. Laboratory work shows that carbon compounds can be better conductors than copper. A whole power system could be built without even touching the rich resources of metal buried in garbage dumps.

Carbon can make windows, of plastic or diamond. Carbon can make things colourful with organic dyes. Carbon can be used to build nanocomputers, and will be the chief component of high-performance nanomachines of all kinds. The other components in all these materials are hydrogen, nitrogen, and oxygen, all found in air and water. Other elements are useful, but seldom necessary. Traces would often be ample.

With a new technology base making recycling easy, there need be no steady depletion of Earth's resources, just to keep a civilization running. The sketch just made shows that recycling just one form of garbage-excess atmospheric CO2-can provide most needs. Even 10 billion wealthy people would not need to strip the Earth of resources. They could make do with what we've already dug up and thrown away, and they wouldn't even need all of that. In short, a twenty-first-century civilization with a population of 10 billion could maintain a high standard of living using nothing but waste from twentieth-century industry, supplemented with modest amounts of air, water, and sunlight. This won't necessarily happen, yet the very fact that it is possible gives a better sense of what the new technology base can mean for the relationship between humanity, resources, and the Earth.

Green Products

In The Green Consumer, Elkington, Hailes, and Makower define a green product as one that:

- Is not dangerous to the health of people or animals
- Does not cause damage to the environment during manufacture, use, or disposal
- Does not consume a disproportionate amount of energy and other resources during manufacture, use, or disposal
- Does not cause unnecessary waste, due either to excessive packaging or to a short useful life
- Does not involve the unnecessary use of or cruelty to animals
- Does not use materials derived from threatened species or environments
- Ideally, does not trade price, quality, nutrition, or convenience for environmental quality

With its ability to make almost anything at low cost-including products designed for extreme safety, durability, efficiency-without mining, logging, harming animals or environments, or producing toxic wastes, molecular manufacturing will make possible greener products than any yet seen in a store. Nanotechnology can replace dirty wealth with green wealth.

The Great Problem

A central problem in environmental restoration is reversing environmental encroachment. We tend to see land as being gobbled up by housing, because the land where we live generally is. Farming, though, consumes more land, and the variant of farming called "forestry" consumes still more. By rolling back our requirement for farmland, and for wood and paper, nanotechnology can change the balance of forces behind environmental encroachment. This should make it more practical, politically and economically, for people to move toward environmental restoration.

Restoring the environment means returning land to what it was-removing what has been added and, where possible, replacing what has been lost. We've seen how this can be done, in part, by removing pollutants and some of the pressures for ploughing and paving. A more difficult problem, though, is restoring the ecological balance where the changes have been biological. Much of Earth's biological diversity has been a result of biological isolation, of islands, seas, mountains, and continents. This isolation has been breached, and reversing the resulting problems is one of the greatest challenges in healing the biosphere.

Imported Species

Human meddling with life in the biosphere has caused enormous ecological disruptions. This hasn't involved genetic engineering-by twisting organisms to better serve human purposes, genetic engineering usually leaves them less able to serve their own purposes, less able to survive and reproduce in the wild. The great disruptions have come from a different source: from globe-travelling human beings taking aggressive, well-adapted species from one part of the planet to another, landing them on a distant island or continent to invade an ecosystem with no evolved defenses. This has happened again and again.

Australia is a classic case. It had been isolated long enough to evolve its own peculiar species quite unfamiliar elsewhere: kangaroos, koalas, duck-billed platypuses. When humans arrived, they brought new species. Whoever brought the first rabbits could not have guessed that they, of all creatures, would be so destructive. They soon overran the continent, destroying crops and grazing lands, unchecked by natural competitors or predators. They were joined by invaders from the plant kingdom: the prickly pear, and others.

The Americas have suffered invasions, too: tumbleweed, a bane of the rancher and farmer, is a relatively recent import from Central Asia. Since 1956, Africanized bees have been spreading from Brazil and moving north-but what they displace, in America, are European bees. Africa, in turn, is being invaded by the American screw-worm fly, an insect with larvae that enter an animal's wounds, including the umbilical wound of a newborn, and eat it alive. The story goes on and on. People have sometimes tried, with a measure of success, to fight fire with fire:

to bring in parasitic species and diseases to attack the imported species and keep its growth within some reasonable bounds. Australia's problem with prickly pear was tackled using an insect from Argentina; the rabbits were cut back-with mixed results-using a viral disease called myxomatosis: "rabbit pox."

Environment Restorement

The 1970s saw a revolution in Western attitudes toward the natural environment. Concern with pollution, deforestation, and species extinction exploded. With the rise of these concerns came an ambivalent attitude toward technology and the wealth it was producing: some said that human beings are destructive to the environment in direct proportion to their power. This immediately suggested that technology and higher living standards were bad, being inherently destructive. "Wealth" came to imply environmental destruction.

The revolution in attitudes toward the environment has changed the idea of wealth. Our national statistics may not reflect it-not every last citizen or politician may agree-but the concept that genuine wealth includes not just houses and refrigerators, factories and machines, cars and roads, but also fields and forests, owls and wolves, clean air, clean water, and wilderness has taken deep root in minds and in politics. "The wealth of nature" has come to include nature as a value in itself, not merely as potential lumber, ore, and farmland.

As a consequence, greater wealth has begun to mean cleaner wealth, greener wealth. Richer countries can afford more expensive, more efficient equipment-scrubbers on smokestacks, catalytic converters on cars-and so they can produce goods with less environmental impact. This trend gives at best a hint of the future.

Lester Milbrath, director of the Research Program in Environment and Society at the State University of New York at Buffalo, observes, "Nanotechnologies have the potential to produce plentiful consumer goods with much lower throughput of materials and much less production of waste, thus reducing carbon dioxide buildup and reducing global warming.

They also have the potential to reduce waste, especially hazardous waste, converting it to natural materials which do not threaten life." James Lovelock states that "The future could be good if we regain a sense of purpose and embrace the new industries based on information and nanotechnology. These add enormous value to molecular-sized pieces of matter, and need not be a threat to the environment as were the heavy polluting industries of the past."

Easier to be Clean

Should we boast of high technology while industry still can't produce without polluting? Pollution is a sign of low technology, of inadequate control of how matter is handled. Inferior goods and hazardous wastes are two sides of one problem.

With processes based on molecular manufacturing, industries will produce superior goods, and by virtue of the same advance in control, will have no need of burning, oiling, washing with solvents and acids, and flushing noxious chemicals down their drains. Molecular-manufacturing processes will rearrange atoms in controlled ways, and can neatly package any unwanted atoms for recycling or return to their source. This intrinsic cleanliness inspired environmentalist Terence McKenna, writing in the Whole Earth Review, to call nanotechnology "the most radical of the green visions."

This green vision will not be fulfilled automatically, but only with effort. Any powerful technology can be used for good or ill, and nanotechnology is no exception. Today, we see scattered progress in environmental cleanup and restoration, some slowing of ecological destruction, because of organised political pressure buoyed by a groundswell of public concern. Yet for all its force, this pressure is spread desperately thin, fighting enormous resistance rooted in economic forces. But if these economic forces vanish, the opposition will crumble. Often, the key to success in battle is to give one's opponents an attractive alternative to fighting. The most powerful cry of the antigreen opposition has been that clearing and polluting the land offer the only path to wealth, the only escape from poverty. Now we can see a clean, efficient, and unobtrusive alternative: green wealth, compatible with natural wealth.

Technologies that will Change the World

For years the public has viewed Nanotechnology as the stuff of Science Fiction — small robots a millionth of an inch in size, capable of transforming the very molecular structure of matter. Whether it's Neal Stephenson's SF novel "The Diamond Age" describing nano-probes building entire cities from just the raw materials, or Start Trek's Borg transforming people into cyborg drones, nanotechnology has appeared to the public as "way out there." However, nanotech has a reality in scientific circles, and a number of companies are starting to reap financial rewards from being early innovators.

Physicist Richard Feynman first suggested the concept of nanotechnology in 1959. In a famous speech he first suggested that devices and materials could be assembled atom-by atom. The field has been nurtured and evangelized by K. Eric Drexler of the Foresight Institute, ever since his 1981 journal article on molecular nanotechnology. The ability to build things on a molecular level- down to nanometre size — will require a whole new set of technologies, as dealing with the ultra-small involves a different set of rules. However, once an "Assembler" which can build things atom-by-atom has become a reality, almost any device that could be conceived (designed and programmed) could be built cheaply from raw materials.

Nanometre devices are not as far away as you might think. Scientists and engineers have learned to create micron-scale "micro-tech" devices of about a thousandth of an inch in size. These MicroElectroMechanical Systems (MEMS) can be built with very small moving parts, and is real, commercial technology today. Techniques for working at a micron level with silicon semiconductors can create ultra-fast computing components, ultra-small sensors and telemetry devices, and ultra-high bandwidth communications networks. They will transform the Internet, revolutionize medicine, and bring visions of the future into reality.

We will have three pioneering companies present their visions of how these technologies will reshape the 21st century, and how they plan to profit from them.

Texas Instruments is a pioneer in the MEMS field with their DLP and DMD (Digital Light Processing and Digital Micromirror Device) technology. Integrating a million controllable mirrors or more on a semiconductor chip, each device is capable of directly converting digital electronic information into dramatic visual images. Today the transport, as well as the display, of high bandwidth information is increasingly done in the photonic domain. MEMS technology has broad applicability to the delivery of high bandwidth content.

Ball Semiconductor has not only invented the techniques for producing semiconductors on small three-dimensional surfaces, they have incorporated MEMS and assembler-like technologies into their plant. On one end of the fabrication line sand is turned into millimetre-sized silicon spheres. Travelling from process to process in plastic tubing, out the other end of the fab will come powerful semiconductor devices.

Zyvex is betting on the future of nanotechnology. Doing basic research into nanotech assembly techniques, their stated goal is to develop the first true Assembler. It may be ten years before their first commercial success, but they have already emerged as the world's first (and perhaps only) commercial nanotechnology company.

History of Nanotechnology

- In a talk given in 1959, Richard Feynman was the first scientist to suggest that devices and materials could someday be fabricated to atomic specifications: "The principles of physics, as far as I can see, do not speak against the possibility of maneuvering things atom by atom."
- A short history of the idea of nanotechnology is given in "Nanotechnology: Evolution of the Concept," Journal of the British Interplanetary Society, Vol. 45, pp. 395-400. This essay was reprinted in the book Prospects in Nanotechnology: Toward Molecular Manufacturing.

Advancing Beneficial Nanotechnology

Foresight is the leading think tank and public interest institute on nanotechnology. Founded in 1986, Foresight was the first organisation to educate society about the benefits and risks of nanotechnology. At that time, nanotechnology was a little-known concept.

Today, with the basic framework of public understanding in place, we are refocusing our efforts on guiding nanotechnology research, public policy and education to address the critical challenges facing humanity. Foresight's new mission is to ensure the beneficial implementation of nanotechnology.

Foresight is accomplishing this by providing balanced, accurate and timely information to help society understand and utilise nanotechnology through public policy activities, publications, guidelines, networking events, tutorials, conferences, roadmaps and prizes.

Foresight is a member-supported organisation. Our membership, including over 14,000 individuals and a growing number of corporations, is diverse demographically and geographically. They are interested in ensuring that the future of nanotechnology unfolds for the benefit of all. These concerned individuals include scientists, engineers, business people, investors, publishers, artists, ethicists, policy makers, interested laypersons, and students from grammar school to graduate level.

Sci-Fi in the Packaging Industry

Nanotechnology is the newest craze among media, academia, investors (just think of the ROI if you hit a company that comes up wit the best "micromachine" for God knows what) and all kinds of industry. How will it affect the packaging industry?

Just a little background on Nanotechnology for those of you who are not familiar with it, do not read comic books or watch the Sci-Fi channel often. Nanoscience and Nanotechnology's primary push, excitement or "umph", if you will, is in the arenas of materials science, electronics, optoelectronics and biomedical. Biomedical is extremely exciting as it can be used hopefully for repairing organs, removing disease, or even making organs operate more efficiently. Then this will move into the military arena. How? A soldier can be repaired faster if injured. Or maybe the soldier can run faster if Nanotechnology or Micromachines are being used to enhance reflexes. The list goes on and on; you can use your imagination.

So Nanotechnology is the ability to manipulate molecules and atoms to create structures, which may be used in the real world. So how does this apply to packaging? Well the science will help deliver materials with greater functionality or durability for the increase in shelf life. Examples, that popcorn bag that junior tries to open will not spill over your new carpet; the coke can that can stand up to punctures a little better. How about that box in your garage that had the cement mix in it, which is now soaked with water? Having to deal with a mess like that would make anyone mad.

Nanotechnology will provide better fracture hardness for aluminum. Note the above example. Nanoscience will provide better tensile strength for items that use carbon fibre and improve flame resistance for plastics with "nanoclay" composites.

Gas-barrier characteristics of Nanoclay in food packaging have sparked great interest. Using Nanoclay in packaging films, i.e. shrink-wrap helps to create a better oxygen barrier. Nanoclay will also be used for the tracking of products through the supply chain. In the next 5 years 5 million pounds of nanocomposite material will go into rigid and flexible packaging. This will affect the packaging of soft drinks, beer, meats and a vast array of packaged foods and condiments. These will be the first consumable items to be packaged in Nanotechnology Packaging. One can certainly hope it will allow one to get that last bit of ketchup out of that little pack whilst one is sitting at the ballpark!!!

One of the problems with Nanotechnology as with any new technology is the cost. This is something new and many companies in the paper, packaging and printing industries cannot really see what Nanotechnology can do for them. Will the cost be outweighed by the benefit, hopefully?

If the affects of Nanotechnology and its challenges are not addressed then the future of compositeness amongst companies and organisations will be threatened.

The application of Nanotechnology shows a lot of promise in the world of packaging. It will vastly improve the packaging materials properties. The only problem is these improvements are years away. Will companies see the benefit of continued investment? Or will they decide not to reinvent the wheel. Who knows? Time will tell.

Technologist Eric Drexler envisioned a future in which machines far smaller than dust motes would construct everything from chairs to rocket engines, atom by atom; in which microscopic robots would heal human ills, cell by cell. Sixteen years after the publication of Drexler's book Engines of Creation, the molecular-scale technologies most immediately available to consumers are somewhat less fantastic: stain-resistant khakis and more durable tennis balls.

Much of the hype is gone from nanotechnology, the term Drexler popularized for his world of very small wonders. But something more interesting has crept in: sales. The khakis and tennis balls are bringing in money, as are dozens of other new products made and enhanced through nanotechnology. To be sure, most nanotech companies are still investing more in R. and D. than they are collecting in revenue. But many commercial applications are in advanced stages of development or already on sale: handheld devices that can sense anthrax spores, hand cream that can protect us from them and computer chips that are faster, cheaper and cooler (we're talking temperature here, not hipness) and retain data even when the power is shut off. Says Richard Smalley, a Rice University professor and Nobel-prizewinning chemist: "We are only beginning to see the things nanotechnology can do."

Nanotechnology takes its name from a nanometre (nm), a billionth of a metre, or about one one-hundred-thousandth the diameter of a human hair. In common usage, it refers to an array of new machines and materials whose key parts are smaller than 100 nanometres and to the new tools, such as Veeco Instruments' atomic-force microscopes and Nanometrics' inspection tools for semiconductor makers, that allow the tiny parts and particles to be observed and manipulated.

It is a mysterious realm in which the laws of classical physics yield to those of quantum mechanics, in which the powerful bonds between atoms overtake the effects of gravity that rule the big world. Yet scientists have moved beyond the basic exploration of nanotech to its exploitation. The National Science Foundation foresees a $1 trillion market by 2015 for nano products, and businesses and governments around the world are rushing to cash in.

The White House has proposed that $710 million be spent on nanotech research next year- a 17% increase over the 2002 budget- on the development of everything from water-filtration equipment to military uniforms made from "smart" materials that can guard against germ warfare. Governments in Asia and Europe are investing $2 billion in similar R. and D., according to CMP-Cientifica, a research firm in Madrid. "Nanotech is a three-legged race right now," says Mark Modzelewski, executive director of the NanoBusiness Alliance, based in New York City.

Even enthusiasts like Modzelewski caution that no one should expect an overnight nanotech revolution. The technology will evolve— "radically," he says- as its benefits seep into virtually every crevice of human industry, from toys to tanks. And even professional investors are cautious. "True venture capitalists are not investing. They are watching," says Glenn Fishbine, author of The Investor's Guide to Nanotechnology and Micromachines. Only a handful of "pure play" nanotech stocks exist, including Nanophase Technologies, in Romeoville, Ill., which makes nanoscale powders, among them zinc oxide particles for sunscreen that won't turn lifeguards' noses white. Still, investors in behemoths such as Intel, Samsung and Dupont have been indirectly funding nanotech development for years.

Five sectors surveyed here- consumer goods, computers, pharmaceuticals, energy and cars- deserve particular attention for the progress they have made toward bringing profitable products to market.

Tires and Toys

Physicist Harris Goldberg wants to revolutionize the $1 billion tire-sealant business, but until that goal is realised, he will settle for tennis balls. InMat, Goldberg's seven-employee company in Hillsborough, N.J., regularly ships to Wilson Sporting Goods 55-gal. drums filled with an environmentally safe liquid containing 1-nm-thick sheets of clay. When the material coats the inside of a tennis ball, it traps air far more effectively than standard rubber alone and doubles the life of the ball. Wilson's Double Core, which made its debut more than a year ago, sells at a premium in U.S. tennis shops and this year became the official ball of the Davis Cup competition.

InMat will take in just $250,000 this year, but Goldberg expects to double that figure in each of the next few years, largely on tennis-ball business. Meanwhile, he is working to convince tire manufacturers that by sealing their wheels with his technology instead of butyl rubber, the current sealant, they can produce tires that run cooler and safer, are lighter and increase a car's fuel efficiency. The U.S. Army has asked InMat to develop gloves that will protect soldiers from chemical agents. Goldberg's funding has come mostly from an angel investor and grants, which means he is still on the prowl for cash. "It's been a battle," he says. "It's still a battle. But we're looking at enormous growth prospects."

A chemical process that adds "nano-whiskers" to cotton fabrics and renders them wrinkle and stain resistant explains new products from Eddie Bauer and Lee Jeans. The fabrics were developed by Nano-Tex, a Greensboro, N.C., company that is 51% owned by Burlington Industries, a textile firm that is struggling to emerge from bankruptcy. Nano-Tex has also developed active-wear fabrics that disperse and dry sweat. Later this year, it will launch a line, destined for socks, underwear and T shirts, that will channel body odors through the structure of the fibres.

Rod MacGregor, a high-tech entrepreneur, runs NanoMuscle, an Antioch, Calif., company that makes 3-in. motors suitable for everything from power windows to dolls with nuanced facial expressions. "I like to be on the wave of the next insanely great thing," he says. His motors work because the alloy nitinol can assume different shapes as its temperature fluctuates. An electrical current causes a nitinol wire in the device to shorten, allowing the linear motor to contract like a human muscle but at 1,000 times the strength.

That's a simple task but an important one, and one MacGregor believes can reach markets worth $3.8 billion. The NanoMuscle, which costs less than $1 to make, qualifies as nanotech, the company says, because of the size of its nitinol crystals, not the wire or motion. MacGregor compares his product to $40-to-$100 small motors made by potential competitor RMB, of Biel-Bienne, Switzerland. Hasbro, a major investor in MacGregor's start-up, expects to deliver its first nano-powered toys by Christmas 2003. NanoMuscle's challenge, like InMat's, will be to stay afloat long enough to sign companies on as clients.

Smart Soot

Like MacGregor, Greg Schmergel is a serial entrepreneur. If his company, Nantero, in Woburn, Mass., is successful, it will eventually add about five minutes to everyone's day. The company wants to build an "instant-on" computer that doesn't need to boot up.

In a year, Nantero expects to produce a commercial prototype for a chip with "nonvolatile random-access memory" (NRAM), which means its chips won't forget how to run all its programs when the power is switched off. The technology uses arrays of 2-nm strands of carbon atoms, called carbon nanotubes, that convey electrons faster than copper and are 100 times as strong as steel at a fraction of the weight. Pairs of tubes store data by locking together when a current runs through them and stay together even when the computer power is switched off and back on. The tubes remain linked until separated by a countercurrent, so their memory is retained. And these chips have other advantages. Schmergel says that within three years, Nantero can bring to market chips with nram that can store 10 times as much data as a silicon chip the same size while operating faster and with less heat. "They're not saying much publicly about their approach," says Steven Glapa, president of the nano-consulting firm In Realis, "but what they're promising sounds pretty breathtaking."

Nanotubes could be the first commodity in the nanotech economy. Dozens of companies around the world already pump out mounds of the stuff- affectionately called soot- and sell it to some of the world's largest companies and labs for research: IBM, Hewlett-Packard, Samsung and NEC. Nano-Lab, in Brighton, Mass., is one of the few nanotech companies turning a profit. It sold $200,000 worth of made-to-order nanotubes in 2001 and is on track to more than double that amount this year. Last week HP researchers unveiled a way of manufacturing molecular-scale circuitry that will be cheaper and use less power than current silicon chips and have the potential to store entire libraries of information.

Cancer Busters

Professional athletes won't find Richard Smalley's soccer balls quite as novel as Harris Goldberg's tennis balls. In fact, without an atomic-force microscope, they won't find them at all: the naturally occurring structures are composed of just 60 carbon atoms. Yet Smalley's discovery is expected to help treat aids, cancer and Lou Gehrig's disease, and it earned him and two colleagues the 1996 Nobel Prize for Chemistry.

These molecular structures are called fullerenes, or buckyballs, in honor of the American architect and inventor Buckminster Fuller. Smalley sits on the board of C-Sixty, a biotech company that builds fullerenes into molecules that researchers hope will attach to and deactivate hiv molecules and blow up cancer cells on cue. "Buckyballs are not quite like nanosubmarines that target deadly diseases"—as seen in the 1966 film Fantastic Voyage—"but because of their size and shape, they are well suited for drug discovery," says Stephen Wilson, co-founder of C-Sixty, based in Houston.

C-Sixty's hypothesis is that buckyballs offer a master-key approach, functioning as a universal molecule that can be, in a sense, weaponized to attack any enzyme or receptor that plays a role in a disease's development. C-Sixty is assembling libraries of new buckyball-based molecules that it will test for potential therapeutic value. Early next year, it will conduct human trials of fullerene-based drugs for hiv and Lou Gehrig's disease. With about one-tenth the toxicity of the current hiv drug cocktails, the company's molecule targets new strains of the constantly mutating virus that are no longer susceptible to treatment. In the case of Lou Gehrig's disease, a degenerative nerve illness, the drug prevents or repairs neurological damage. C-Sixty has no revenue yet but will soon announce a partnership with "one of the top three pharmaceutical companies," says its president, Uri Sagman.

Other researchers are building nanoinventions so small that they can slip inside diseased cells and halt their development. Dr. James Baker, head of the University of Michigan's Centre for Biologic Nanotechnology, is developing nanoscale molecules called dendrimers to target cancer cells. He says that "nanotechnology gives us a totally new set of tools to diagnose and treat disease," including an ability to eliminate cells before they become cancerous.

Baker and his team created a company called NanoBio. An $11 million Pentagon grant allowed the team to develop a cream that can penetrate and kill infectious microbes, everything from the fungus that grows on toenails to flu viruses to anthrax spores. The military version, called NanoDefend, is a liquid designed to decontaminate clothing and surfaces that have come into contact with anthrax, Ebola or smallpox. A creamy gel or goop, called NanoGreen, can be used by the military to decontaminate skin- and may eventually have topical and vaginal applications for consumers, according to NanoBio ceo Ted Annis. The firm, which hopes to partner with existing companies, is preparing to submit seven products to the fda for approval, a process that takes several years.

Some companies will simply help common drugs work more efficiently. Elan Drug Delivery, located in King of Prussia, Pa., pulverizes existing drugs to a size that maximizes the body's ability to absorb them. Naproxen sodium, a pain medication found in products such as Aleve, can take as long as two hours to exert its pain-relieving effect. Nano Systems has developed a crystal version of naproxen, still in clinical development, that works in 15 to 20 minutes. "Using NanoCrystals has not made naproxen a better drug"—just seven to eight times as fast as the commercial product, says Larry Sternson, president of drug delivery at Elan, the Dublin-based drugmaker that is the parent company of Elan Drug Delivery.

Detection and analysis are also enhanced by small technology that is not strictly nano-scale. MesoSystems, a young but profitable firm, sells to fire departments handheld devices that collect biological particles

0.5 to 10 microns across- anthrax, for one- and preserve them in a liquid for identification. MesoSystems supplies Lockheed Martin with an air sampler it uses in its Biomail Solutions product, a biohazard detector in field testing at some federal agencies. MesoSystems made about $250,000 last year on revenues of $7 million and this year hopes to gross more than $10 million.

Cleaner Energy

As an alternative to fossil fuel, everyone loves hydrogen fuel cells, which produce clean energy out of hydrogen and oxygen. But hydrogen, while abundant in the air, isn't widely available in refined form. And machines that run on hydrogen are equally scarce. Researchers at the Tokyo Institute of Technology have been working on the first problem, automakers on the second. The Tokyo group has developed a way to "crack" hydrogen, using a mesh of thin carbon fibres studded with molecules of a nickel compound. The filter breaks down natural gas into carbon and hydrogen that is pure enough for use in fuel cells. Another impediment is the cost and supply of the platinum particles that catalyze, or kick off, the process. Think of them almost as matchmakers, encouraging every oxygen atom to mate with two hydrogens, releasing valuable energy with each reaction. That is the heart of the fuel cell.

Because of the current size of these catalyst particles, about 10 nm, and their tendency to clump together, platinum is not used efficiently. The world's entire annual output of platinum would not meet the demand if fuel cells were used by only 10% of cars produced worldwide. Hydrocarbon Technologies- which is owned by Headwaters, an alternative-energy company based in Draper, Utah- says it has found a way to create nanoscale platinum particles that won't clump together and slow down the process, as current ones do. The new particles are expected to keep fuel cells running in a stable, efficient manner and stretch the platinum supply. Tim Harper, founder of CMP-Cientifica, says these particles show how "nanotechnology can make previously uneconomic processes viable" for businesses.

OPEC has more than fuel cells to worry about from nanotechnology. Last month China's largest coal company licensed U.S. technology that will enable it to build a $2 billion coal-liquefaction plant in Inner Mongolia. The heart of this new technology is a gel-based nanoscale catalyst that improves the efficiency of coal conversion and reduces the cost of producing clean transportation fuels. If the technology lives up to its promise and can economically transform coal into diesel fuel and gasoline, coal-rich countries such as the U.S., China and Germany could depend far less on imported oil. At the same time, acid-rain pollution would be reduced because the liquefaction strips coal of harmful sulfur. Given current world oil prices (about $27 per bbl.), turning coal into gas is economical in China. "A $4-to-$8-per-bbl. increase in the price of oil would make it economically attractive in the U.S. too," says Theo Lee, ceo of Hydrocarbon Technologies.

Plastic Cars

Every big carmaker promises that any year now, it will have a fuel-cell car on the road- a vehicle that will cruise silently, spit drinkable water from its tail pipe and provide power to your house when you plug it into the garage. In the meantime, auto manufacturers are putting nanotechnology to work in other ways. Toyota was the first to experiment with strong, lightweight nanocomposite materials in the late 1980s, and U.S. automakers are starting to move nanocomposites out of the lab and into vehicles. General Motors is using advanced plastics to make step assists for 2002 GMC Safari and Chevrolet Astro vans.

The new materials are stiffer, lighter and less brittle in cold temperatures than other plastics. Improvements in strength and reductions in weight lead to fuel savings. The next step is for GM to use nanocomposites in car interiors and bumpers and eventually in load-bearing structural parts, such as vehicle frames.

As nanotechnology produces more products and processes, will the technology ever catch up with Eric Drexler's theories? Says Steve Bent, a Washington patent lawyer for nanotech firms: "That will be the research agenda for the rest of the century."

Tools of the Trade

A primary long-term goal of nanotechnology research now underway worldwide is the creation of micro machines-mobile, versatile devices and systems that can perform engineering tasks or medical procedures on a microscopic scale.

Thanks to their minuscule size and motility, microdevices will open up new frontiers in molecular materials and manufacturing, electronics, health care, medical research, biotechnology, and information technology. Before micromachines become a reality, however, researchers must learn a lot more about the properties and interactions of materials in the micro- world, where forces other than gravity dominate. They'll also have to develop the tools and technology for building micromachines, which are the focus of research efforts underway at the University's Advanced Microsystems Laboratory.

Researchers there already have developed smart microassembly systems, atomic-level force sensors, and various microdevices to aid the development of micromachines, says Bradley Nelson, associate professor of mechanical engineering.

Some of these parts are no larger than one one-hundredth of the width of a human hair, says Nelson, who investigates the physical interactions among micromechanical parts during the assembly process. Lab researchers had to build special sensors that could measure nanonewtons of surface-effect forces over nanometres of separation between parts.

"We're working to develop a new manufacturing paradigm so these parts can be assembled economically into intelligent microscopic machines," he explains.

Micromechatronic systems span a broad range of applications, from microsurgical tools for minimally invasive surgery to machines that can manipulate bits of information and improve disk-drive density to "clouds of intelligent dust" that monitor the environment. Other applications for Nelson's research include the automatic assembly, computer vision, manufacturing, and robotics industries. Another University researcher is studying active thin films now being developed for small-scale sensors and micromachines. Mathematics professor Mitchell Luskin is collaborating with colleagues from several IT departments to understand the behavior of thin films of martensitic, ferroelectric, and magnetostrictive materials. Luskin is working to apply this research to the development of microelectromechanical systems (MEMS) that can have an important impact on biomedicine.

"A transformation is just beginning in biomedicine that promises to revolutionize how some surgeries and therapies are done," Luskin says. MEMS may someday replace the knife and scalpel as tools for minimally invasive surgery and also deliver powerful drugs directly to targeted areas in the body.

According to Nelson, the University has emerged as a national leader in the research and development of nanotechnology and micromachines. "The University is a real player in this field," he says.

Nanotechnology in the Present World

Nanotechnology is a science in its infancy, but that doesn't mean it hasn't been put to use. What exactly has been accomplished in nanotechnology so far? In general, all of today's practical nanotechnologies are those using nano-size particles of various materials, or nanometre-size features on integrated circuits (ICs), rather than the complex molecular machines that engineers first envisioned. These current nanotechnologies are still made by "top down" methods (like those used in conventional chemistry and IC manufacturing), rather than the largely unproven "bottom up" techniques predicted by nanotechnology's boosters.

Many current nanotechnologies, for example, consist of the ever-shrinking transistors, interconnecting wires, and other features on digital ICs. As of 2005, some integrated circuits now have transistors that measure about 50 nanometres across-well inside the accepted size-based definition of nanotechnology. But chips are still made using advanced versions of the lithographic processes developed in the 1950s, which layer on materials and then carve away at them to form the electronic circuits. They are not, in other words, constructed molecule-by-molecule from the bottom up. However, chip manufacturers point out that when working with extremely small circuit elements, the behavior of electrons changes, so entirely new principles are at work. Also, there is at least one new chip with a somewhat different claim to being "nanotechnological." This is IBM's "Millipede" memory chip, which draws its inspiration directly from the Atomic Force Microscope (AFM). Electronics manufacturers can also point to the latest generation of high-density computer hard drives, which have extremely thin coatings of just a few atoms' thickness applied to the surface of the disc by a process called chemical vapor deposition.

While such nanochips are beginning to appear in greater numbers, most of us more often encounter applications of nanotechnological materials that are made in "bulk" form and added to other products. By far the best known of these are the controversial "nanotechnology" trousers introduced by The Gap and Eddie Bauer stores in 2005. These were simply ordinary cotton pants, treated with nanoparticles of a new, stain-resistant chemical that attached itself to the cotton molecules.

Carbon nanotubes, which can now be made in large quantities at relatively low cost by companies like Hyperion Catalysis International Inc., are being incorporated into a wide range of other products. Because the fibres conduct electricity very well, Hyperion was able to mix them into plastic compounds, which auto makers can then mold into parts that conduct electricity. This is useful for preventing static electricity charges from building up on parts such as plastic fuel system components, where the static can eventually damage them or, in some cases, cause a spark.

Nanotubes mixed into plastics are very strong and light, and have been used to make car body components, tennis rackets, and other items. They have also been used to improve battery performance, and may some day be used in other technologies that traditionally used ordinary carbon or metals to conduct charges. Infineon Technologies in Germany, for example, has demonstrated the use of the tubes to connect components on microchips. In 2002 they showed how nanotubes could be used to replace ordinary metal wires allowing them to carry more current but taking up less space. That would result in computer chips that can pack more circuits into less space; one of the longstanding goals of chip designers.

One very useful new material is the semiconductor quantum dot. While not used in electronic circuits, quantum dots are nonetheless made from the same silicon used in computer chips. These tiny bits of material are coming into widespread use in experimental biology and, in a limited way, in medical diagnosis. The dots can be coated with certain chemicals, which are specially formulated so that they bind themselves to particular things-such as RNA, cell walls, or other types of molecules found in cells. One interesting application of this technology is its use in analyzing DNA material taken from the body. These DNA "scanners," first introduced commercially by Matsushita Corporation, combine integrated circuit technology and quantum dots to analyze genetic material much more rapidly than was possible before, and may lead to more rapid assessment of diseases.

A second use of coated quantum dots is injecting them into the body, where they circulate until they come in contact with whatever type of cell their coating is designed to attach itself to. Then when a powerful infrared light source is shone on the body, it penetrates the flesh, illuminates the massed quantum dots, and the reflections can be detected to provide a "live" picture of an organ, muscle, cancerous growth, or other internal part without the need for surgery. Unfortunately, not all of these quantum dots are suitable for injection into a living human body, and some are even poisonous, but bioengineers are working around that problem.

Even with these real-world applications, the current uses of nanotechnology (other than nano-size particles of various materials) remain very limited. In fact, several once-promising nanotechnology based systems introduced commercially in the 1990s did not meet with success, such as the nanotube-based Field Emission Displays proposed as competitors to other flat-panel information displays. However, researchers are rapidly making progress toward what some think of as true nanotechnologies-self-assembling, molecule size machines to perform all sorts of tasks (including manufacturing the nano-size materials made by other methods today). The nanotechnological future, we are told, is right around the corner.

The Future of Nanotechnology

The future of nanotechnology is largely a question mark. Futurists say we are entering a new era, somewhat like the Industrial Revolution of the 18th and 19th centuries. That revolution changed nearly everything about the way people lived. But no one at that time could have predicted how those changes would unfold. Could we be on the brink of another very rapid period of profound technological and social change?

The nanotechnological revolution, if it occurs, will be just as unpredictable in the long-term, but scientists and engineers have laid out some pretty fantastic forecasts for the near-future. For example, some see great promise for the use of nanotubes in super-strong materials. Even though the plastic composites made today using relatively short nanotubes are not yet much stronger than earlier types of composites, long nanotubes are expected to be used for extraordinary applications like the proposed "space elevator." This system would replace rockets for the transport of payloads and people into earth orbit.

Another major area where nanotechnologists predict stunning changes is in medicine. Imagine a world where no one gets seriously ill, grows older, or even dies (until they want to). That is what the prophets of nanotechnology say is in store for the 21st century. Today's nanotechnologies used in medicine offer only modest benefits, such as the ability to target diseased or cancerous cells, making them easier to locate.

In the near future, engineers tell us, that will change. Tiny molecular machines, perhaps based on complex, branched molecules called "dendrimers" will be injected into the body not only to locate cancers but also to find and repair cells damaged by disease or aging. Livers and hearts damaged by natural wear-and-tear, inherited diseases, poor nutrition, or alcoholism will be fixed or even replaced. Genetically based ailments such as Alzheimer's will be cured by replacing the faulty genes.

Some futurists have predicted that the most profound changes will be the result of the introduction of molecular assembly "factories," perahps even small enough to fit on a desktop. These would, some say, make it possible for virtually anyone to design and build virtually anything, using nanorobots or perhaps a new technology called "nanoink," created by NanoInk founder Chad Mirkin. Nanotechnology researchers like K. Eric Drexler and Ralph Merkle say that this could be done by imitating and improving upon the "manufacturing" that DNA accomplishes inside the body. In 1999, researchers such as Nadrian Seeman at New York University demonstrated the principle of using modified DNA molecules to build a tinymachine, and somewhat later Nanoink founder Chad Mirkin had demonstrated building up nanostructures by depositing layers of materials on a substrate. These and other experiments have also led researchers to believe that they will eventually be able to assemble circuits atom-by-atom in order to create the next generation of computer chips. The circuits on these chips will be much smaller than what is currently possible, and will enable the building of much more powerful computers. What difference will that make? Some, like engineer Raymond Kurzweil, think that computers will have personalities and be as smart as humans within 20 years. We may even be able to "download" our own personalities into computers, to become virtual humans. With nearly unlimited computing power, programmers are sure they could create software that completely blows away anything possible today.

Not surprising, these amazing predictions have inspired fear as well as wonder. Environmentalists and others point out that nanotechnology may bring with it unexpected dangers. The nanomaterials being made today, like fullerenes, are often in the form of extremely small particles. Even when these particles are made from common materials like carbon, they may interact with the human body or the environment in ways that are unlike those of natural particles of the same materials. Some say that allowing nanoparticles to be included in products ingested or applied to the body may pose health risks for consumers.

Others predict that nanotechnology may get out of control, causing a huge man-made disaster. Eric Drexler and others, such as computer engineer Bill Joy of Sun Microsystems, warned in 2000 that self-replicating machines might run amok if they escape into the environment, competing with natural bacteria, plants, and people for natural resources. Then, in 2002 the public's awareness of nanotechnology-the bad side of nanotechnology-was greatly expanded when author Michael Crichton published his best-selling novel Prey, about tiny, self-duplicating nanorobots that band together to try to take over the world. Whether public fears are founded in fact, it is true that the future of nanotechnology has inspired as much caution as optimism. Recently, in response to public outcry, researchers such as Dr. Vicky Colvin of Rice University have begun evaluating the risks and rewards of current nanotechnologies. Colvin and other engineers believe that, with wisdom, they can bring the wonders of nanotechnology into being while avoiding the pitfalls.

Nanotechnology to Pave Way for Micro-Machines

Disposable satellite transmitters, inexpensive medical testing equipment and sensors for automatically tracking inventory or traffic patterns will become possible over the next 10 years through developments in nanotechnology, speakers at the Nanotech 2003 conference said Monday.

Nanotechnology—the science of making devices with features measuring less than 100 nanometres (or one-ten-millionth of a metre)—will let companies make smaller and cheaper products than ever before, which in turn will lead to new markets, according to Albert Pisano, a professor of engineering and computer science at the University of California at Berkeley.

Communications, for instance, could be greatly affected, Pisano noted. Right now, radios on satellites have to be hardened against radiation, an expensive process. At Stanford University, researchers have shown how small, unhardened radios can transmit and receive messages while in orbit. Although these smaller radios are subject to harm from radiation, they cost far less, so many could be mounted on a spacecraft to compensate for burnouts.

A market for tiny transmitters could also emerge within five years, out of a need to better monitor power lines, freeways and bridges for potential failures.

The movement will also mark a major turning point in mechanical engineering because most of these tiny systems will deal directly with the physical world, like pumps do, rather than getting information from electrical impulses, like microprocessors.

"Most people forget that a radio is two-thirds filter and that filters are mechanical," Pisano said. "Sensors and computing and communications can all be heavily miniaturized."

Batteries, too, could improve through breakthroughs in mechanical engineering. Currently, a kilogram of lithium ion, the same material used in notebook batteries, provides about 400 watt-hours of energy. A kilogram of methane provides 15,000 watt-hours of energy. The difference is that lithium ion is electrically charged, while methane provides thermal energy—that is, heat.

"Can you use thermal energy on such a small scale?" Pisano asked. Apparently so. Some researchers have already developed rotary engines that measure only microns across. Of course, getting to miniature nirvana won't be easy. In the early '90s, futurists predicted a booming market for MEMS (microelectromechanical systems) similar to the devices described above. Difficulties in mass manufacturing, packaging and other problems kept the market from taking flight.

Getting There

Pisano, though, noted that new technologies take time. About 15 years passed between the development of the transistor in the late '40s to the true development of the electronics industry. Several challenges need to be overcome, but the progress with nanotechnology makes MEMS more promising.

The first must-have applications for the types of MEMS products Pisano described will appear in about five years, he predicted, because the demand exists. Gillette, for instance, has spent heavily in the last few months on diminutive radio-frequency transmitters.

A number of universities and private companies will be presenting papers on MEMS over the five-day conference here. Some of the major themes include microfluidics, or small machines that can test biological samples, and films or surfaces with sensory capabilities.

Other companies are presenting papers on progress in using chains of molecules to make processors or memory devices. Molecular computing will likely take much longer to develop, but early results are showing promise.

Molecular chips differ substantially from today's silicon processors, noted Phaedon Avouris, manager of nanoscience and nanotechnology at IBM's Watson Research Centre. Carbon nanotubes exposed to the air, for instance, carry a positive charge. When hermetically sealed off, they carry a negative charge.

Nanotech's Small Truth

The drop-off was stomach-churning. In 2000, venture capitalists poured $100 billion into startups. Last year, they couldn't even reach $40 billion. So forgive them for latching onto nanotech as the uptrend. Headlines like "NANOTECHNOLOGY WINS OVER MAINSTREAM VENTURE CAPITALISTS" and "THE NEXT BIG THING IS VERY SMALL" are getting hard to avoid. Nano conferences are weekly events, crowded with VCs amped up about self-assembling machines and nanobots in your bloodstream.

But what's really going on? While venture capitalists are happy to hype nanotech, they aren't exactly rushing in where angel investors fear to tread. Of the $5.6 billion put up by private capital sources in the first three months of this year, only $42 million went to very small tech, according to Steve Glapa, president of In Realis, a market consulting firm in Milpitas, California. Even putting all of these investments under the "nanotech" umbrella is a stretch. Advanced chemistry and materials science are more like it. The reality of nanotechnology has less to do with self-assembling Ferraris than with powders and filters and goop. It's about taking a material like zinc oxide and grinding it down to grains the size of a few dozen nanometres (great for sunscreen). Or punching nano-size holes in screens so hydrogen atoms can slip through but water molecules can't (handy for fuel cells). Or laying down slurries a few atoms thick (good for hard drives). It's not a new industry so much as the outcome of decades of research. "People have been doing this sort of thing for a long time," says Glapa. "We just didn't always attach the nano prefix to it."

So why all the fuss from VCs now? Why the conferences, the speeches, the sound bites? Because it's great business. Attaching nano to a company name can instantly multiply the value. Or at least that's the idea. "VCs invest in companies where they see an exit strategy," says Glenn Fishbine, author of The Investor's Guide to Nanotechnology and Micromachines. "The more they can do to make nanotechnology attractive now, the better off they'll be when it's time to sell."

Unlike past technology booms, it's unclear whether VCs will ever have much of a role to play in the nanotech realm. Industrial labs at IBM, Hewlett-Packard, and 3M have invested hundreds of millions of dollars in scaling materials down to the atomic level, a gap that VCs are unlikely to close. "They are playing in exactly the same market," says Tom Theis, IBM's director of materials research. "They have no advantage over what big companies have looked at and rejected."

The structure of VC investing- seed money, ramp-up, IPO, exit-won't work easily with nanotech. It's an area where $800,000 government grants are readily available, often enough for a research scientist to get to a working prototype. The next step, bringing a product to market, will often take longer than VCs are willing to wait. So researchers with promising technologies are likely to bypass VCs altogether and sell directly to an established industrial player. That's the route Nano-Tex took when the Emeryville, California-based company developed a nanoscale technology to protect cotton from stains, then sold it to Burlington. "If someone has a really good memory technology, they won't go to a VC, they'll just go to IBM," says Fishbine. "Then they don't have to worry about marketing and manufacturing and distribution; they'll just take the royalty payments and buy land on Kauai."

Not so fast, says Steve Jurvetson of Valley VC firm Draper Fisher Jurvetson. Before startups can partner with big industrial players or license intellectual property to them, they need backing from venture capitalists. At the same time, he adds, VCs need to relearn patience. During the Internet craze, the moneybags on Sand Hill Road got used to going from concept to IPO in as little as 18 months. Post-bust, VCs are returning to the standard three- to seven-year wait for a return. “I think we’re on the long end of that for nanotech,” says Jurvetson, adding that a dozen years isn’t out of the question. “If the opportunity is big enough, if it’s really going to change the world in a significant way, then it’s worth the wait.”

That thinking- and time horizon- may work for today’s nano of chemistry and goop. But the VCs are luring investors with a grander vision: the idea that someday we might manipulate matter in fantastic ways, that we could build complex structures from the atomic level-much the way living things self-assemble from info stored in their DNA. Of course, that’s all speculative stuff, and even nano evangelists like Richard Smalley say we’re decades from realising that dream, well beyond even Jurvetson’s expanded VC timeline. So don’t believe the hype. The next great nanotech triumph is not going to be about visionary investment mavericks who see the truth long before the dull, slow multinationals. It’s going to be about someone slogging away in a basement lab for 15 years. This time around, the deep-pocketed corporations will lead the charge. To do this story justice, we’re going to need a brand-new script. Less about the money, more about the science. Less sizzle, more steak.

Sustainable Architecture

Sustainable architecture is architecture that seeks to minimize the negative environmental impact of buildings by efficiency and moderation in the use of materials, energy, and development space. Sustainable architecture uses a conscious approach to energy and ecological conservation in the design of the built environment.

The idea of sustainability, or ecological design, is to ensure that our actions and decisions today do not inhibit the opportunities of future generations.

Sustainable Energy Use

Energy efficiency over the entire life cycle of a building is the single most important goal of sustainable architecture. Architects use many different techniques to reduce the energy needs of buildings and increase their ability to capture or generate their own energy.

Heating, Ventilation and Cooling System Efficiency

The most important and cost-effective element of an efficient heating, ventilating, and air conditioning (HVAC) system is a well-insulated building. A more efficient building requires less heat generating or dissipating power, but may require more ventilation capacity to expel polluted indoor air.

Significant amounts of energy are flushed out of buildings in the water, air and compost streams. Off the shelf, on-site energy recycling technologies can effectively recapture energy from waste hot water and stale air and transfer that energy into incoming fresh cold water or fresh air. Recapture of energy for uses other than gardening from compost leaving buildings requires centralized anaerobic digesters.

HVAC systems are powered by motors. Copper, versus other metal conductors, helps to improve the electrical energy efficiencies of motors, thereby enhancing the sustainability of electrical building components.

Site and building orientation have some major effects on a building's HVAC efficiency.

Passive solar building design allows buildings to harness the energy of the sun efficiently without the use of any active solar mechanisms such as photovoltaic cells or solar hot water panels. Typically passive solar building designs incorporate materials with high thermal mass that retain heat effectively and strong insulation that works to prevent heat escape. Low energy designs also requires the use of solar shading, by means of awnings, blinds or shutters, to relieve the solar heat gain in summer and to reduce the need for artificial cooling. In addition, low energy buildings typically have a very low surface area to volume ratio to minimize heat loss. This means that sprawling multi-winged building designs (often thought to look more "organic") are often avoided in favour of more centralized structures. Traditional cold climate buildings such as American colonial saltbox designs provide a good historical model for centralized heat efficiency in a small-scale building.

Windows are placed to maximize the input of heat-creating light while minimizing the loss of heat through glass, a poor insulator. In the northern hemisphere this usually involves installing a large number of south-facing windows to collect direct sun and severely restricting the number of north-facing windows. Certain window types, such as double or triple glazed insulated windows with gas filled spaces and low emissivity (low-E) coatings, provide much better insulation than single-pane glass windows. Preventing excess solar gain by means of solar shading devices in the summer months is important to reduce cooling needs. Deciduous trees are often planted in front of windows to block excessive sun in summer with their leaves but allow light through in winter when their leaves fall off. Louvers or light shelves are installed to allow the sunlight in during the winter (when the sun is lower in the sky) and keep it out in the summer (when the sun is high in the sky). Coniferous or evergreen plants are often planted to the north of buildings to shield against cold north winds.

In colder climates, heating systems are a primary focus for sustainable architecture because they are typically one of the largest single energy drains in buildings.

In warmer climates where cooling is a primary concern, passive solar designs can also be very effective. Masonry building materials with high thermal mass are very valuable for retaining the cool temperatures of night throughout the day. In addition builders often opt for sprawling single story structures in order to maximize surface area and heat loss. Buildings are often designed to capture and channel existing winds, particularly the especially cool winds coming from nearby bodies of water. Many of these valuable strategies are employed in some way by the traditional architecture of warm regions, such as south-western mission buildings.

In climates with four seasons, an integrated energy system will increase in efficiency: when the building is well insulated, when it is sited to work with the forces of nature, when heat is recaptured (to be used immediately or stored), when the heat plant relying on fossil fuels or electricity is greater than 100% efficient, and when renewable energy is used.

Bibliography

Altmann, J. & Gubrud, M.A.: *Discovering the Nanoscale*, IOS Press, Amsterdam, 2004.

Andrew, N.: *Foundations of Nanomechanics: from Solid-state Theory to Device Applications*, Berlin, New York, 2003.

Attebery, B.: *Nanoculture: Implications of the New Technoscience*, Intellect Books, Bristol, 2004.

Bainbridge, W.S. : *Societal Implications of Nanoscience and Nanotechnology*, Dordrecht, Kluwer, 2001.

Bast, R.C. Jr.: *Cancer Medicine*, Decker, Hamilton, 2000.

Berube, D.M.: *Discovering the Nanoscale*, IOS Press, Amsterdam, 2004.

Brosnan, J.: *The Primal Scream: A History of Science Fiction Film*, Orbit Books, London, 1991.

Carroll, J.S.: *Social Science Research Methods for Assessing Societal Implications of Nanotechnology*, Dordrecht, Kluwer, 2001.

Caviedes, C.: *El Nino in History: Storming Through the Ages*, University Press of Florida, Florida, 2001.

Credi, A.: *Molecular Devices and Machines: a Journey into the Nano World*, Wiley-VCH, Weinheim, 2003.

Crow, M. & Sarewitz, D.: *Societal Implications of Nanoscience and Nanotechnology*, Dordrecht: Kluwer, 2001.

Daniel Ratner: *Nanotechnology: a Gentle Introduction to the Next Big Idea*, Upper Saddle River, Prentice Hall, 2003.

Dobzhanshy, T.: *Genetics of the Evolutionary Process*, Columbia University Press, New York, 1970.

Drexler, K. Eric: *Engines of Creation: The Coming Era of Nanotechnology*, Anchor Books, New York, 1986.

Dyson, Tim: *India's Historical Demography: Studies in Famine, Disease, and Society*, Riverdale, Westwood, MA, 1989.

Eamonn Gearon: *The Sahara: A Cultural History*, Signal Books, UK, 2011.

Frank J. Owens: *Introduction to Nanotechnology*, Hoboken, J. Wiley, 2003.

Glimell, H.: *Dynamics of the Emerging Field of Nanoscience*, Dordrecht, Kluwer, 2001.

Glimell, H.: *Societal Implications of Nanoscience and Nanotechnology*, Kluwer, Dordrecht, 2001.

Haber, L. F.: *The Chemical Industry: 1900-1930, International Growth and Technological Change*, Oxford: Claredon Press, 1971.

Hessenbruch, A.: *Nanotechnology and the Negotiation of Novelty*, Amsterdam, IOS Press, 2004.

James Lewis: *Prospects in Nanotechnology: Toward Molecular Manufacturing,* John Wiley & Sons, New York, 1995.

Karn, Barbara: *Nanotechnology and the Environment: Applications and Implications,* American Chemical Society, Washington D.C., 2004.

Keblinski, P. : *Encyclopedia of Nanoscience and Nanotechnology*, American Scientific, Los Angeles, Calif, USA, 2004.

Liu W.K., Karpov E.G., Park H.S.: *Nano Mechanics and Materials,* Wiley, Weinheim, 2006.

Mayr, E.: *Change of Genetic Environment and Evolution*, Allen and Unwin, London, 1954.

Mody, C.C.M.: *How Probe Microscopists Became Nanotechnologists, Discovering the Nanoscale*, IOS Press, Amsterdam, 2004.

Norio Taniguchi: *Nanotechnology: Integrated Processing Systems for Ultra-precision and Ultra-fine Products,* Oxford University Press, New York, 1996.

Orth, R. N.: *Nanometer Scale Antibody Patterning for Directed cell Immobilization and Stimulation,* Cornell University, Ithaca, 2002.

Pressman, J.: *Nano Narrative: A Parable from Electronic Literature*, Intellect Books, Bristol, 2004.

Pressman, J.: *Nanoculture: Implications of the New Technoscience*, Intellect Books, Bristol, UK, 2004.

Ratner, Daniel: *Nanotechnology: a Gentle Introduction to the Next Big Idea,* Prentice Hall, Upper Saddle River, 2003.

Robison, W.L.: *Discovering the Nanoscale*, IOS Press, Amsterdam, 2004.

Sarewitz, D.: *Nanotechnology and Societal Transformation*, Dordrecht, Kluwer, 2001.

Stephen M. Goodnick: *Transport in Nanostructures,* Cambridge University Press, Cambridge, 1997.

Tenner, E.: *Nanotechnology and Unintended Consequences,* , Dordrecht, Kluwer, 2001.

Vasudev Bhatnagar: *Cell Science and Technology*, Campus Books, Delhi, 2009.

Vermaas, P.E.: *Discovering the Nanoscale*, IOS Press, Amsterdam, 2004.

Weil, V.: *Ethical Issues in Nanotechnology*, Dordrecht, Kluwer, 2001.

Index

I

L

M

N

O

P

R

S

T

U

W

❑❑❑